Build Your Own Wireless LAN

Build Your Own Wireless LAN

James Trulove

McGraw-Hill
New York Chicago San Francisco Lisbon
London Madrid Mexico City Milan New Delhi
San Juan Seoul Singapore Sydney Toronto

McGraw-Hill

A Division of The ***McGraw-Hill*** *Companies*

1 2 3 4 5 6 7 8 9 0 DOC/DOC 0 9 8 7 6 5 4 3 2

ISBN 0-07-138045-0

The sponsoring editor for this book was Majorie Spencer, the editing supervisor was Steven Melvin, and the production supervisor was Sherri Souffrance. It was set in New Century Schoolbook by Patricia Wallenburg.

Printed and bound by R. R. Donnelley & Sons Company.

This book is printed on recycled, acid-free paper containing a minimum of 50 percent recycled, de-inked fiber.

CONTENTS

Tech Tips

ACKNOWLEDGMENTS

Many people have helped with the process of creating this book. We would particularly like to acknowledge the assistance of Malcolm Myler and Paul Hines in providing wireless LAN equipment and case studies; Mike Wolleben of Wi-LAN for providing technical information; Joe Thiele N5SMN for valuable insights in 2.4 GHz operation and compatibility; Marjorie Spencer, brilliant acquisitions editor at McGraw-Hill, and her able editorial assistant, Jessica Hornick; Patricia Wallenburg of TypeWriting for flawless and rapid book production; and finally for Ruth Ann who helped enormously with the entire writing process, including thorough proofreading of the final copy. In addition, kudos are due to Alfred Bolinger, for his absolutely flawless PrintKey freeware program that gives the Print Screen button the functionality for which it was intended, and a good deal more. PrintKey is available from shareware and freeware sites worldwide.

James Trulove
jtrulove@BuildYourOwnWirelessLAN.com

CHAPTER 1

Wireless Revolution

Chapter Highlights

- Freedom from Wires
- WLANs Are Radio Devices
- WLANs Are Ethernet
- Benefits of WLANs
- Wireless Standards
- IEEE 802.11 and WiFi
- Chapter by Chapter

For most of the past 50 years of the Computer Age, digital computer equipment has been tethered to the walls of our offices and homes by cables. Computer technology has gradually advanced from bulky equipment to compact personal computers that have 100, 1000, even 10,000 times the capacity and speed of their predecessors. Until recently, the need for a power cable to our desktop PC made the presence of another cable for our connection to a modem or a local area network (LAN) fairly trivial, if not logical. However, the advent of portable devices, such as laptop computers and hand-held personal digital assistants (PDAs), has brought home the inconvenience of a data cable in this wired world.

Freedom from Wires

Laptop and hand-held computers give us a considerable amount of freedom. We can take our computer applications with us to the office conference room, the café, or the easy chair at home. And, if we used our computers only for so-called personal productivity applications, such as word processing and spreadsheets, we could be totally disconnected from the world. Increasingly, however, we are using these portable computers to access the World Wide Web and the Internet. A connection to a local network or remote access to an Internet service provider is required to reach the Web. What is the use of a battery-powered computer if we still have to run a data cable to the wall?

Now, we can access our network wirelessly! That is, we can add a simple adapter card to our portable computer that will make a wireless connection to a LAN. This will let us connect to the Internet to get e-mail or to browse the Web for information, entertainment, or commerce. We can also use various LAN resources, such as printers and fax machines, from our wireless connection.

Wireless networking gives us an incredible amount of freedom and flexibility. We can get connected almost anywhere. The locations where casual wireless network access is available are rapidly multiplying. Home and small-office wireless networking kits are widely available and relatively inexpensive. Wireless networking is here, and this book gives you the information you need to take advantage of this exciting technology.

The wireless revolution has begun.

Wireless Is Everywhere

The term *wireless* refers to the transmission of information without wires. This is generally understood to mean transmission via radio waves, that is, *electromagnetic propagation*. A sea of radio waves already surrounds us with a wide variety of information content. For example, FM radio and broadcast television are transmitted to our homes, cars, and portable devices. Radio transmissions to pagers use similar techniques, but with different frequencies. Digital pagers can receive text messages, which are wireless data transmissions. Two-way transmissions are also common. We use cellular telephones and two-way pagers in our daily lives. These are wireless devices that connect us to our communication services and, hence, to our world.

To understand the nature of radio transmission, we need to understand the characteristics of these familiar wireless services. Then we can use some of the intuitive knowledge all of us have regarding radio and TV operations to help us understand how to implement wireless data networks. The principles are essentially the same. Only the names have been changed to protect the innocent.

WLANs Are Radio Devices

Actually, the most obvious difference between wireless LAN (WLAN) transmission and the familiar radio and TV transmissions is the *frequency*.[1] The frequency of a radio signal is the rate of change, in Hertz (Hz; or cycles per second), of an electromagnetic signal. For these radio and television transmissions, the rate of change occurs millions of times per second and is thus expressed in megahertz.

Many things in nature have frequency as a component. For example, voices are conveyed through sound waves in a range of frequencies from about 300 to 3000 Hz. Musically, the note A-above-middle-C on the piano is 440 Hz. We can convert these audible sound waves to electrical signals with microphones (such as those in telephones) and then convert them back to sound waves with transducers such as earphones and speakers. We call the range of frequencies that correspond to these audi-

---

[1]A second fundamental difference is that data, rather than voice and video, is transmitted. A third difference is the way in which the information is placed on the frequency channel. WLANs use a modulation technique called *spread spectrum* rather than conventional amplitude or frequency modulation.

ble sound waves *audio frequencies*. Audio frequencies can be transmitted over wires, of course, but they can also be transmitted wirelessly as electromagnetic waves, although with some difficulty.[2]

Frequencies that are considerably higher than the audio frequencies were used initially for the transmission of radio broadcasts and are consequently called *radio frequencies* (RF). Figure 1.1 shows some common frequencies in many day-to-day encounters with RF transmission. Because RF occurs at many thousands or millions of hertz, we use the magnitude prefixes of *kilo*, *mega*, and *giga* for thousands, millions, and billions, respectively. We abbreviate the respective frequencies as kHz, MHz, and GHz. Table 1.1 shows a listing of how these abbreviations translate into real terms.

Figure 1.1 The frequency spectrum.

Audio Frequencies	300 Hz	Voice and
	3000 Hz	Music
	20,000 Hz	Percussion–Cymbals
Radio Frequencies	530 kHz	
		AM Broadcast Radio
HF-Range	1630 kHz	
	1800 kHz	
		Short-wave Radio
	30 MHZ	
VHF-Range	54–88 MHz	TV Channels 2–6
	88–108 MHz	FM Broadcast Radio
UHF-Range	450 MHz	UHF TV Broadcast
	800 MHz	Two-way Radio
	850 MHz	Cellular Radio
	900 MHz	Pagers and Two-way Radio
Microwave	1 GHz	Long-range Radar
ISM Band→	2.4 GHz	Wireless LANs and Phones
UNII Band→	5.3–5.8 GHz	Wireless LANs, etc.

[2]We transmit very low frequency signals, near the audio range, to submarines at sea because those frequencies travel long distances in the earth and penetrate the ocean to a far greater depth than higher frequencies. Omega, a predecessor to modern global positioning systems, used high audio frequencies that were transmitted as radio waves from ground stations around the world.

TABLE 1.1
Numerical Terms for Magnitudes Used in Networking

Unit	Symbol	Example	Example Symbol	Order of Magnitude	Numeric Equivalent
kilo	k	kilohertz kilobits/second	kHz kbps	10^3	1,000
mega	M	megahertz megabits/second	MHz Mbps	10^6	1,000,000
giga	G	gigahertz gigabits/second	GHz Gbps	10^9	1,000,000,000
tera	T	terabyte	TB	10^{12}	1,000,000,000,000
milli	m	milliwatts	mW	10^{-3}	1/1,000
micro	μ	microvolts	μV	10^{-6}	1/1,000,000

Wireless—Back to the Future!

In the early days of RF transmission, the use of telegraph lines was replaced by what was called "wireless" transmission. The name caught on because it was quite appropriate—transmission without wires. When voice broadcasts began, early radios and their transmissions were often referred to as "the wireless." The advent of television and modern entertainment radio allowed the term to fall into history—for a while. It is said that history often repeats itself, and this is true of the term *wireless*. Again, we have a very logical use in data transmission. Data has traditionally been transmitted between computers and along local networks as a wired technology. Accordingly, when we started to replace these wires and cables with electromagnetic transmission, the term *wireless* gained a new and thoroughly modern meaning.

In the context of WLANs, we use the terms *radio* and *wireless* somewhat interchangeably. This is natural because both are terms for the same underlying technology. In some cases, the term *wireless* actually seems sort of magical, which it is not. However, we do not ordinarily "experience" wireless data connections, so exactly how it works is a mystery to most people. Obviously, radio and TV are integral parts of our lives. Indeed, we often intuitively understand radio transmission because of our direct experiences. After all, we know our cell phones work poorly in the hills and mountains and not at all inside areas of certain buildings. We also have noticed that AM radio stations disappear while we drive through a tunnel or underpass. These are characteristics of WLAN operation as well, so an understanding of ordinary radio transmission and reception is invaluable in gaining insight into WLAN operation.

We describe wireless networking as a radio (electromagnetic) phenomenon. However, infrared transmission also can be wireless. Some of the WLAN standards actually include infrared networking standards, but we will stick to the radio transmission here.

WLANs are being rapidly implemented everywhere. More and more businesses and public venues are adding WLANs to their services.

Wireless Networking and Wireless Access

In this book, we focus primarily on wireless networking, specifically the WLAN connection as defined by the IEEE 802.11 standard. As the name implies, a *wireless* LAN is a wireless connection to a LAN, as shown in Figure 1.2. A LAN, in today's terms, is simply a direct Ethernet connection between two or more *nodes*, such as two PCs. Actually, an Ethernet node can be a PC, a server, a router, a workstation, a cable modem, or, as we will see, a wireless access point. The important point is that it is a *local* connection.

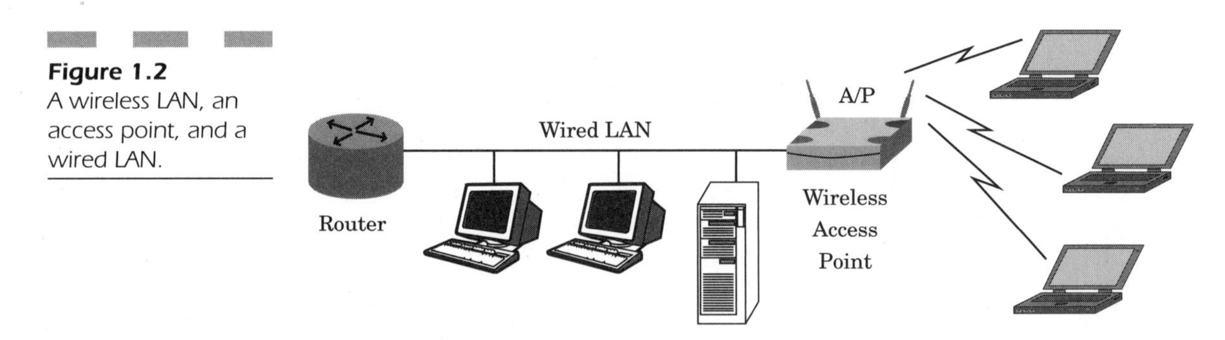

Figure 1.2 A wireless LAN, an access point, and a wired LAN.

[3]As shown in the Tech Tip on Long Reach LAN Connections, long-reach fiber optic Ethernet links are possible up to more than 40 miles, but the signal remains in native format, rather than being encapsulated in a wide-area protocol. The reader is referred to *LAN Wiring*, 2nd ed., by James Trulove (McGraw-Hill, 2000) for detailed information on wired LANs.

In a wired Ethernet network, local literally means local. Network (specifically LAN) connections are allowed to be 100 meters (329 feet) over copper cables, or somewhat farther over fiber.[3] The WLAN is similarly local, with typical open-air operating distances of 100 to 500 meters (300 to 1500 feet).[4] A WLAN may consist of one or more devices (nodes) that are connected over a radio link to a wireless access point (AP). The AP functions as a bridge between the WLAN and the wired LAN.

Tech Tip: Long Reach LAN Connections

Technically, we can run some of these so-called LAN connections over substantial distances, but they remain electronically in the same native signal format as the shorter versions. For example, WLANs can operate point to point at distances longer than 10 miles with special equipment. A special fiber-optic version of the Gigabit Ethernet standard can extend more than 40 miles. In both cases, however, the data structure stays the same as in the shorter-distance connections. The standard short-range fiber Gigabit Ethernet link is called SX, or more properly, 1000Base-SX. A moderate-range link called LX is available and can go up to 5000 m, or about 3 miles. The long-reach Gigabit Ethernet connection is called ZX and can go as far as 70 km.

WLANs Are Still Ethernet

The fundamental idea to remember about a WLAN or a wired LAN connection is that the signaling structure of the Ethernet data packet does not have to change. The WLAN connection is what we use to interconnect our computers and other portable devices to a conventional wired LAN. A distinguishing feature of the WLAN is its operation at near "wired" speeds. We can easily make WLAN connections from 11 to 54 Mbps (and eventually faster than 100 Mbps).

We also have a type of connection called *wireless access*. Wireless access allows us to connect portable devices, primarily cell phones and personal digital assistants (PDAs), to an Internet service provider, typically over

[4]A WLAN link can be extended by the use of high-gain antennas and amplifiers to more than 10 miles, as we will see later. It remains in native format and is not encapsulated to do this.

the cellular telephone network. Wireless access may also be used to connect to desktop computers in areas where high-speed lines are not available. Wireless access speeds generally range from 19.2 to 128 kbps. That's as much as 500 times slower than WLANs!

With wireless access, we convert the data, if it is already in Ethernet format, and encapsulate it in a form that can be sent to a central site. Normally, wireless access gives us an Internet connection, so that we can get our low data-rate services such as e-mail and stock quotes. With the very low speed of most wireless access networks, we really have to keep the amount of data to a minimum.

Here is a simple analogy of the conversion that takes place with wireless access. As you know, we can easily talk to each other, even yell to each other, over relatively short distances. The "signal" is our voice, and it is carried over sound waves. Early ships even used sound tubes so the captain could talk to the engine room. This is similar to the LAN (except that we would be "sending" much more information per second on the LAN). However, to go great distances, we must use a telephone. The telephone converts our voice to an electrical signal that can be amplified, digitized, and encapsulated in a high-speed network, and carried halfway across the planet—or to the moon. The telephone circuit is similar to wireless access, as it connects at low speeds, but over long distances, and coincidentally uses the voice network to make the connection.

Features and Benefits of Wireless Networking

A WLAN extends the coverage of a LAN into the surrounding area without the need for wires or cables. This means that your network connection is totally portable without having to worry about cumbersome and inconvenient wires and cables. In general, you can jump on to any available local network, with some minor adjustments. Of course, there may be some additional steps for data security, but the basic connection is quite simple.

Wireless Standards Are the Key

WLANs currently operate in the 2.4-GHz and 5.2–5.8-GHz frequency bands. Both frequency ranges are covered in the IEEE 802.11 standards. The very popular 2.4-GHz WLANs are actually in a subsection of the standard, called 802.11b. This technology expanded the operating speed of the WLAN from 2 to 11 Mbps. An enhancement of the standard is afoot that increases the speed to 54 Mbps. The actual speed you will get from a real-world 11-Mbps installation will vary up to about 6 Mbps throughput. As you may know, even a wired Ethernet network typically peaks at about 50 to 60 percent use so we are talking about the same throughput as a 10-Mbps LAN.

The 5-GHz WLAN band is standardized, rather out of logical order, in IEEE 802.11a. It is a much newer technology and not yet widely used. It offers higher data rates, which will eventually reach to near Fast Ethernet speeds. In its initial release, 802.11a supported a variety of speeds up to 54 Mbps.

If you buy WLAN equipment, you will most likely see the IEEE 802.11b operation featured prominently because there are several older technologies, some of which were proprietary to a particular manufacturer. You will want to choose the IEEE 802.11b (or perhaps the IEEE 802.11a) models for greatest compatibility.

WiFi and Wireless Compatibility

An industry group, the Wireless Ethernet Compatibility Alliance (WECA) sponsors the WiFi™ Program for wireless fidelity. The group tests and certifies WLAN equipment for compatibility and interoperability. As you can appreciate, it is critical that all this equipment operate seamlessly together despite the different chip sets, software, and manufacturing techniques. This group has been instrumental in ensuring proper interoperation of WLAN devices. It is important to look for the WiFi logo on WLAN gear.

Distance Capability of WLANs

WLANs are low-power, local, wireless networks. That means that the distances will range from less than 100 m (~300 feet) to no more than 500 m (~1500 feet). The exact distance will depend on several factors, such as walls, windows, and interference sources. The output power on these WLANs is quite low for the most part. One reason the output power is limited is to minimize interference between systems. You wouldn't want someone's WLAN in the next block to interfere with yours. Consequently, the distance coverage is fairly short. After all, it is a "local" wireless network.

TABLE 1.2
WLAN Characteristics

Frequency Bands	2.4 GHz or 5.2–5.8 GHz
Channels	Up to 11
Distance	300–1500 feet (100–500 m)
Data Rates	1, 2, 5.5, 11, 22* Mbps (2.4 GHz)
	6, 12, 24 Mbps (5.2–5.8 Gbps, mandatory)
	9, 18, 36, 48, 54 Mbps (5.2–5.8 Gbps, optional)
Power Output	100 mW (20 dBm) to 1 W (30 dBm) in the USA

** Future growth to this rate.*

The general characteristics of WLANs are shown in Table 1.2. As the table shows, 802.11b allows the speed to downshift as the signal gets weaker. When the signal strength is weaker, the wireless link automatically adjusts the data rate to compensate. This is much like the way you tend to speak more slowly and distinctly on a bad phone line.

We will cover all of the variations in the WLAN environment that affect usable distance. You should plan your installation carefully to maximize WLAN throughput and distance. As with any computer technology, attention to details will produce much better results. To understand WLANs thoroughly, we will go into greater detail in later chapters.

WLAN Benefits

What are the benefits of this WLAN technology? In an office, workers can carry their portable computers into conference rooms, cafeterias, and even co-workers' offices. Of course, it has always been possible to use a laptop PC in a stand-alone, disconnected manner for word processing and other applications. However, with the WLAN you are now connected to your office network. You can print to a printer, send e-mail, share presentations and other files, and connect to the Internet. You can even conduct a videoconference! The possibilities for increased productivity and collaboration are endless.

At home, you free yourself from sitting at a fixed computer. If you have a portable computer, such as a laptop PC, you can move virtually anywhere in your home. For that matter, you can move anywhere outside your home, at least to the porch or yard. You can connect to a home network or to the Internet. You can even interconnect to your company's private network through a secure connection called a virtual private network (VPN).

Just Imagine...

You can surf the Internet from your easy chair. Using your wireless network, you can sit on the couch or at the kitchen table and still have access to everything—servers, printers, Internet—you had when you were tethered to the wall. Probably, you were previously sitting in a rather uncomfortable chair, squinting at a monitor, and facing that same wall.

Put a little variety in your life. Do your e-mail in the easy chair. If you are like many modern office workers, you have to work at home just to catch up. Many of us use the evening to wade through the volume of e-mail that comes in each day. Imagine doing that while you are sitting in the family room with music or the TV playing. That would certainly take some of the tedium out of this chore.

While you are there, you could browse the Internet. Perhaps something is mentioned on television that you want to know more about. Maybe that vacation villa in Maui? Or, you might want to check out local activities, sports events, concerts, or just scan the news, in print and on the net. Remember, there are no wires, so you can go to any room, go outside, or even relax, propped up in bed.

Outside would be nice. You can work from your deck, or poolside, or just in the chaise lounge chair on the patio. Direct sunlight is tough on laptop displays and your eyes, so you might want to find some shade. You can imagine, can't you, that you could be sitting on the beach? Perhaps at that villa in Maui that you found surfing from your living room.

Traveling and Learning with WLANs

On your way to a business trip (or to Maui) you will have to stop in an airport waiting room. Many airports have added WLANs to their waiting rooms, restaurants, and airline clubs. Imagine you are picking up and responding to your e-mail in the airport. You will also find many hotels and conference centers with WLANs. That's a feature to look for when you check out that villa's Web page. Actually, many hotels will be featuring WLAN connections to help you make the right decision to encourage you stay with them.

If you are enrolled in a college or university, you may be able to use a WLAN to connect to the campus network. Can you imagine doing your necessary class work anywhere on campus? You could go to the library, to the lab, to the classroom, or to the quad (doesn't every university have a quadrangle—that large, grassy area between the major buildings?). Of course, you would want to make that laptop use totally appropriate, never needlessly surfing the Internet during class. Right.

This wireless surfing has become so prevalent at many campuses that you may be asked to keep your laptop shut during class. It is too bad that you can't take electronic notes just because some bozo was caught going somewhere out of bounds. On the other hand, you can always wirelessly surf before and after class, which can save countless quantities of midnight oil.

On a more serious note, we will be covering office and campus WLANs later in this book. It turns out that there is much subtlety in providing wide coverage in these areas. We will show you all the planning and installation tips and tricks you will need to know. Whether you are directly involved in planning such an installation or just a knowledgeable user, you will benefit from this chapter.

What We Will Cover

As you know, this book is about building your own WLAN. In this book, we cover home, office, and even campus wireless networks. We include information about design, installation, and use of WLAN products, from the basic WLAN interface card in your computer to the AP devices that bridge between wired and wireless worlds.

This book is arranged from simple to complex, with clearly set out technical tips and topics of special interest. We concentrate on what you need to *know* to successfully install a WLAN. To that end, we skip over some of the more complicated technical issues in the text of the chapters. However, when there is a really interesting technical topic, or when the devil is in the details, we will insert a boxed text area called a

Tech Tip. We will also include five projects on wireless networking. These are designed to "get specific" about WLANs. Too often technical books lack a practical side. We stay very practical in this book, you can be assured.

We begin with the essentials in Chapter 2, *Wireless LAN Basics*. In this chapter, we describe the components of a wireless network, the interface cards, the access points (essentially a wireless hub), and the antennas. Also, we develop some useful analogies for thinking about WLAN operation. This will help you intuitively plan your WLAN and make troubleshooting and correcting problems much easier. Finally, we will talk about the frequency bands that are used for WLANs and potential interference and coverage issues.

In Chapter 3, *Internet in Every Room*, we talk about our first practical application. We cover all the details of choosing wireless network equipment, the hub, and network interface card (NIC). You are guided through a typical wireless NIC installation and see some samples of the software tools available to help you install and administer your home WLAN. We will see where to place the AP for best coverage of your home, how to use external antennas to boost your range, and several tricks. We get specific about how some very common structures in your house, such as walls, tile, and even air vents, can have an effect on your wireless signal. Finally, we do a project to build a simple wireless home network and interconnect it to the Internet.

Chapter 4, *Home Office and Small Office WLANs*, expands our understanding of WLANs to the home office or small office environment. Here we get to add a small wired LAN and show you how to get wireless access to centralized resources such as printers, fax, file servers, and routers. We talk about the use of cable modems, digital subscriber lines (DSL), and some of the minor technical issues you need to know to get connected—such as dynamic host control protocol (DHCP) and network address translation (NAT). Because it is becoming increasingly common to have more than one PC connected to our high-speed DSL or cable modem lines, it is important to know how to add a WLAN to such a network. The small/home office WLAN project is only slightly more complex than that of the home network, but it shows you exactly how to proceed.

You can get all the details on using your home/office network in Chapter 5, *Virtual Office from Your Easy Chair*. Here we cover the details of connecting your wireless PC to the common resources on your conventional network. We also cover some of the issues with wireless phones, using Internet phone services, and such routine tasks as file backup. The project in this chapter helps you create your virtual office and covers the methods of interconnecting with Microsoft™ Networking.

Chapter 6, *WLAN Security*, gives you information on keeping your private information to yourself. Wireless networking does have some security issues, but technology has been incorporated in the wireless devices to minimize the risk. You will learn about these features and what additional steps you should take. We also discuss more powerful encryption techniques, NAT, and the use of personal firewalls.

A WLAN is not the only way to get wireless advantages in your home. In Chapter 7, *The High-Tech Home*, we talk about very innovative ways to merge wired and wireless home features. We show you how to control your lights, your air conditioning, home security, and even your appliances with your computer—wirelessly, if you want. We also cover some new technologies that can eliminate cables from keyboard, mouse, or PDA. These short-range devices, such as the Bluetooth™ wireless interface, can remove the mess of cables and cords that pour out of the back of every desktop PC, and can make our laptops much easier to use. In addition, they can let appliances, phones, TVs, and sound systems connect cordlessly.

Chapter 8, *Wireless Office and Campus LANs*, covers larger wireless networks. This is a detailed chapter that shows how to plan and implement a large office or campus wireless network. It also shows many of the tips, tricks, and tales of WLANs in these environments. Our project for this chapter is an example of the larger network. We will plan, design, and cover implementation details.

Travel throughout the world with us in Chapter 9, *Traveling Wirelessly*. In this chapter, we talk about connecting in airports and hotels and even in restaurants, schools, and libraries. You get information on choosing your wireless service provider, how to connect, and a review of security issues. If you are involved in providing services in one of these facilities, you will see how to improve range and coverage and how to minimize interference.

Chapter 10, *Internet-enabled Cellular Phones and PDAs* is all about the other type of wireless data networking, wireless access. Cellular telephones and radio-equipped PDAs provide Internet access over special networks that are interconnected to the system of cellular radio towers throughout the countryside. We cover the details of that technology and what to expect from those services.

The Appendix addresses some of the more mundane technical details about wireless networking and how RF technology works. There is information on 802.11 standards and the common frequency bands, channels, and some speed versus distance information. There is simple path model, which includes losses, antenna rules-of-thumb, and typical guidelines for cable loss. All this is reference information to help you plan your wireless network.

Now, let's get started—Build Your Own Wireless LAN!

Wireless LAN Basics

Chapter Highlights

- WLAN Standards
- Wireless NICs
- Ad Hoc and AP Networks
- Thinking about WLANs
- Choosing a W-NIC
- Antennas and Cables
- Interference to WLANs

The first step in building your home or office wireless LAN (WLAN) is to understand how the technology works. In this chapter, we cover the basic concepts you will need to put your wireless network in place. We also discuss a few of the more technical topics, such as propagation, gain, antennas, losses, and interference sources. In addition, we discuss some useful ways of thinking about wireless signals to make the technology more intuitive.

Wireless Just Replaces Wires

If you understand basically how a wired LAN works, you will understand the WLAN. In a conventional wired LAN (Figure 2.1), you must have a network interface card (NIC) installed in your computer, plus a hub or switch to allow you to connect to other local computers, and perhaps a bridge or a router to connect to the Internet. All of these network items are connected by cabling. A wireless network has all of these same components but one. The only item missing from this scene with a wireless network is the network cable.[1]

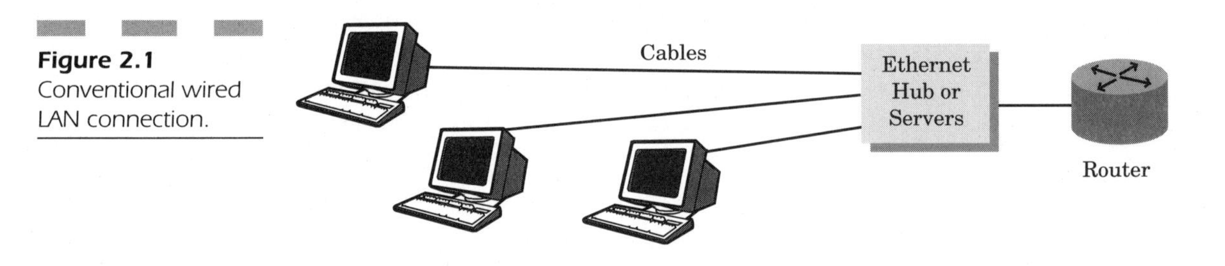

Figure 2.1 Conventional wired LAN connection.

Of course, because a wired LAN is tethered to the walls, the cables from each workstation must first run to a permanently mounted data jack. The data jack, in turn, is permanently wired through a patch panel to an Ethernet hub or switch in some central location (called a telecommunications room). This part of the LAN cabling runs up through the walls, across the ceiling space, and down to the patch panel. As long as the cable remains unbroken, it can run through wood, wallboard, metal

[1]If you are interested in the details of wired LANs and network cabling, check out *LAN Wiring*, 2nd ed., by James Trulove (McGraw-Hill, 2000).

vents, concrete, and even steel. Wired LANs do have the advantage of being able to run around and through obstacles that might block wireless signals.

But, when it is all said and done, you still have the cumbersome wire and cables between you and the hub. WLANs, as we will see, allow you to make the same connection, at comparable speeds, without wires.

The main components of a WLAN (Figure 2.2), include the wireless NIC (W-NIC), the wireless hub or access point (AP), and, if needed, the antennas and cables. As you can see, the WLAN signals go directly to the wireless access point, avoiding the bends and corners of the network cables in the wired LAN. This illustrates the *line-of-sight* nature of wireless networking. This concept is very important in understanding WLAN operation.

Although the components of a WLAN system are analogous to those of the wired system, there are important differences you need to know for a successful WLAN installation. For example, a WLAN connection occurs on one of eleven or so WLAN channels. You must ensure that the wireless NICs are set to the proper channel to connect to the access point. Some W-NICs are frequency agile; that is, they can scan through a range of channels to locate an access point. However, the access point is always set to one particular channel, and the W-NICs may be as well.

As you can see, you must be concerned about a number of additional factors when you install a WLAN. In this chapter, we cover each of those items in some detail. In addition, we help you visualize the WLAN connection and talk about potential sources of interference.

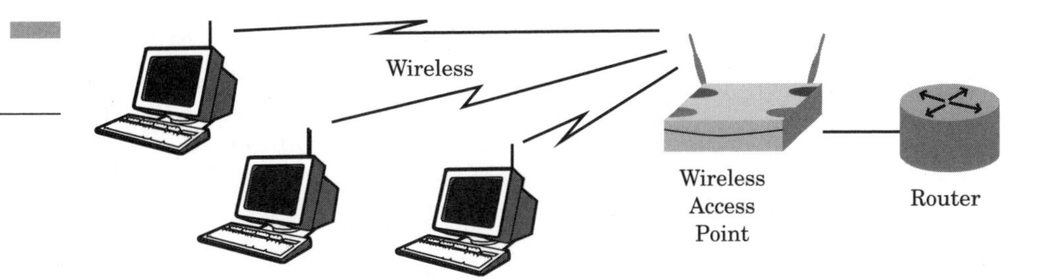

Figure 2.2 Typical WLAN.

Installing a WLAN is really not difficult. If you were to purchase a commercially available WLAN kit and install it in your home or office, it is likely that you would meet with total success at least 50 percent of the time. However, a 50 percent success rate implies a 50 percent failure rate, which is obviously not acceptable. To reach a favorable outcome, you need to have some specialized knowledge regarding wireless networks.

A WLAN uses a type of radio transmission to send network data back and forth between two points. These WLAN transmissions are in a frequency band at 2.4 or 5.2 GHz. At these frequencies and at the low-power levels used by WLANs, the signal is attenuated by walls and ceilings and is easily blocked by metallic objects, such as air ducts or metal doors. In addition, for outside operation, buildings, trees, and even landscaping can attenuate or block the WLAN signal. We need to have an understanding of all of these matters so that we can arrange our network in a way that will allow it to operate properly.

WLAN Standards

Just as wired networks have standards, such as TIA/EIA-568-B and Categories 5e and 6, wireless networks also have standards. The applicable standard for most of the WLAN world is the IEEE 802.11 WLAN standard.[2] In addition to the base standard, there have been several amendments to add new specifications. The amendments of IEEE standards traditionally are referred to by the base standard number, plus the amendment letter, such as 802.11a and 802.11b. Customarily, the new technological features continue to be referred to by that number/letter combination. Periodically the IEEE issues a new base standard document that incorporates all of the approved addenda. However, the industry still refers to the amendment label to highlight compatibility with that particular capability.

Virtually all of the WLAN systems that are currently available have been brought up to the 11-Mbps capability and feature the IEEE 802.11b compatibility prominently in their marketing. You might see "11 Mbps" mentioned, but this also indicates 802.11b compatibility. We should point out that the 11-Mbps rating refers to the maximum data rate of the payload data; that is, the actual information part of the packet you are sending. As shown in the Tech Tip, that is only a portion, albeit a major portion, of the data packet, so you really don't ever

[2]The IEEE-802.11 standard is officially named *Part 11: Wireless LAN Medium Access Control (MAC) and Physical Layer (PHY) Specifications*. It has been adopted by the International Organization for Standardization (ISO) and the International Electrotechnical Commission (IEC) and redesignated as ISO/IEC 8802-11:1999(E). The basic 802.11 document runs well over 500 pages, much of it very detailed technical specifications that are of interest only to product designers. The pertinent major specifications in the standard are mentioned in the Tech Tips.

Tech Tip: The IEEE 802.11 WLAN Standards

The original base specification IEEE 802.11 standard, issued in 1999, specified WLAN operation in the 2.4-GHz ISM band.[3] At the time, WLAN technology was limited to speeds of 1 and 2 Mbps. The direct sequence spread spectrum (DSSS) signaling method was outlined in this standard, as was frequency hopping spread spectrum (FHSS). The preamble and header portions of each packet were sent at 1 Mbps, and the payload data was at 1 or 2 Mbps. However, the data signal was placed on a varying sequence of frequencies (chipped) at an 11-Mbps rate.

In IEEE 802.11b, a new high-rate operating mode was specified at 11 Mbps. A fall-back rate of 5.5 Mbps was also specified because conditions of range and interference could cause the high-rate signal to be unusable. However, two types of packets are possible. In the long preamble/header (HR/DSSS/Long) version, the preamble and header are sent at 1 Mbps. In the short preamble/header (HR/DSSS/Short) version, a shortened preamble (half the length, 72 bits) is sent at 1 Mbps, a normal-length 48 bit header is sent at 2 Mbps, and then the data payload is sent at 2, 5.5, or 11 Mbps.

This allows the wireless link to be adjusted as needed to compensate for signal and interference conditions. The lower speed is used for the preamble, encoding the signal in a highly readable Barker code spreading method with differential binary phase-shift keying (DBPSK) for maximum versatility. The data is then sent at the appropriate rate. This method allows devices to detect transmissions and adjust the payload data rate as needed to compensate for reception conditions.

achieve the full 11-Mbps rate as throughput. As a matter of fact, you usually see 4.5 to 6-Mbps maximum throughput in a typical WLAN, which is comparable to the practical throughput of a conventional wired 10-Mbps Ethernet link.

Figure 2.3 shows all the components in a typical WLAN kit. The kit contains the wireless AP that acts as a bridge between the WLAN and the wired world. In addition, the WLAN kit also contains one or more

[3]The 2.4-GHz ISM band is designated for unlicensed industrial, scientific, and medical purposes in North America and Europe.

W-NICs. The typical home-office kit uses an AP with a built-in WLAN card and antenna. In some cases, the WLAN card function will be incorporated into the AP's circuitry. However, some APs are available with one or two slots for WLAN cards and a provision for external antennas. These more advanced (and more expensive) APs are very useful in complex commercial environments and office buildings.

Figure 2.3 The WLAN kit. Courtesy Linksys.

Another important logo to look for is the WiFi™ mark[4] (Figure 2.4). This indicates that the device has been tested and found to comply with the Wireless Fidelity standards for interoperability. The WECA organization rigorously tests 802.11 equipment from different manufacturers to ensure that the different components can work together. This is similar to the self-testing that many manufacturers do, but, as we all know,

[4]WiFi™, the standard for wireless fidelity, is a certification program administered by the Wireless Ethernet Compatibility Alliance (WECA).

self-certification can be self-serving. Some manufacturers may advertise their equipment as "WiFi Compatible" or "WiFi Compliant," which is not the same as WiFi Certified.

Figure 2.4 The WiFi logo. Courtesy WECA.

Many of us remember the problems that high-speed modem manufacturers had getting their so-called standards-compatible modems to talk together. For a time, you had to make sure that you had the proper brand of V34 or V90 modem to match the ones that the service provider had. Otherwise, you could link up, but you couldn't pass data, at least not at full speed. This was incredibly frustrating until the manufacturers of the chip sets modified their products to work together.

The situation is enormously more complicated in WLANs, because the "standards" permit a lot of variety. Now that there are an increasing number of chip set manufacturers and a larger complement of equipment manufacturers who program those chip sets for their WLAN products, the potential for problems is immense. At last count, there were more than 300 manufacturers of WLAN devices, so you can see the breadth of the issue. The WiFi testing program provides a means for those manufacturers to check their designs and make corrections before we buy their products. Otherwise, you could be stuck with wireless equipment that works only with devices from that same manufacturer. Also, be aware that there are some non-802.11b, non-WiFi products for sale, probably at a big discount now that these standards of operation have emerged so look for that WiFi mark. For more information, you can go to www.wirelessethernet.org, the site of WECA.

Although we are covering primarily 802.11 WLANs in this book, there are several other technologies for wireless networking and cable elimination (see Tech Tip on Wireless Standards). We cover some of these in Chapter 7, The High-tech Home, and elsewhere. The important thing to remember is that 802.11 is the dominant standard at this time, and an incredible amount of equipment is available to implement it.

TECH TIP: WIRELESS STANDARDS

Several wireless networking standards exist for WLAN connections. Here is a list of the most prominent and some of their characteristics.

Standard	Band/Mode	Data Rate	Use
IEEE 802.11	2.4 GHz/DSSS	1, 2, 5.5, 11 Mbps	WLAN
	5 GHz/ODFM	6, 12, 24 Mbps (optional 9 ,18, 36, 48 ,54 Mbps)	WLAN
Home RF™	2.4 GHz/FHSS	800–1600 kbps 8 8 kbps voice	Cable elimination Cordless phones
Bluetooth™	2.4 GHz/FHSS	57–720 kbps 3 64 kbps voice	Cable elimination Cordless phones

Wireless Network Interface Cards

The key component of the WLAN is the wireless network interface card (W-NIC). The W-NIC is the card that connects your computer to a WLAN. The W-NIC contains the RF transceiver circuitry, an antenna, and the PC interface. It is typically a Type II PC card that can fit directly into the appropriate slot in a laptop computer. This type of adapter card was originally known as a "PCMCIA" card, but the term is often shortened to "PC."

The W-NIC is the wireless equivalent of the NIC used with a regular 10/100-Mbps wired network. As with a regular NIC, the W-NIC must be installed in your computer with the appropriate software drivers and configured for proper operation. However, configuration for the W-NIC may have a few additional steps that are required for proper operation in your particular WLAN environment. We will get to these steps in a later section in this chapter.

There are many manufacturers of W-NIC cards, and there are several producers of the wireless chip sets that are incorporated into these cards. The basic functions of the cards are the same, but some may have added features, such as encryption and antenna options. A typical W-NIC was shown in Figure 2.3. The W-NIC looks pretty much like any PC card, except for a boxy protrusion on the end of the card. This plastic covering protects the tiny wireless antenna that sends and receives the wireless signal to and from other units in the wireless network.

W-NIC Antennas

Theoretically, W-NICs come in a variety of shapes and sizes. In actuality, there is not a whole lot of variation, at least in the cards with built-in antennas. To accommodate the antenna, the card simply sticks out a half inch or so (about 15–20 mm) from the side of the laptop computer,when it is installed. A W-NIC installed into a desktop computer typically fits into a standard PCI adapter card such that the antenna box is just outside the rear panel. Some cards have movable antennas, and others have tiny connectors for external antennas. Although these cards do offer more versatility, they typically are more expensive. The antenna options can extend the operating range of the card (that is, the distance you can be from the AP).

You can get greater range with an external antenna, as with any transmitter/receiver combination. Even if you do not have experience with two-way radio communication, you probably have known someone who had a car antenna break. The radio really didn't work very well without a good external antenna. Radios, similarly, do not work well inside large buildings. However, if you move near a window or door, the reception improves dramatically. If you can run an external antenna, performance will increase even more. This may be most evident with a desktop PC, where the case can block the signal.

The tiny antenna on the W-NIC card is a compromise between performance and size. A truly efficient simple antenna would be at least 6 cm (about 2.5 inches) long at these frequencies, which is a little too long for the PC card. So a smaller antenna is used to fit on the card, but the performance is less than optimal. The answer to this problem is to use an external antenna. If your card manufacturer has a provision for an add-on antenna, you can regain the lost range by adding a compact antenna that sits beside the computer. You also can add a larger, directional antenna to greatly extend the workable distance from the AP.

Of course, a bulky or cumbersome external antenna will limit your mobility if you are a laptop user. On the other hand, if you need to extend your wireless coverage to a desktop PC, the added antenna will not be a problem. There are many types of antennas, with signal gains of 3 to 25 dB. Some are highly directional and some have coverage in all directions (omnidirectional). We cover proper selection and use of antennas in Chapters 3 and 8, and a more technical discussion in the Appendix.

Tech Tip: Antenna Size

An RF antenna is more effective at certain mathematical element lengths relative to the wavelength of its design frequency. Effectiveness is judged in terms of antenna pattern, relative gain, and driving impedance (which should match the attached transmitter or receiver). Antennas are normally sized at certain fractional wavelengths, such as $\frac{1}{4}$, $\frac{1}{2}$, or $\frac{5}{8}$.

For example, one very practical length for a simple antenna is $\frac{1}{2}$, wavelength, which is commonly called a *dipole*. The physical length is related to the electromagnetic wavelength, so a dipole would be one-half the wavelength. Thus, a dipole for 2.4 GHz would be 6.25 cm (62.5 mm, or about 2.5 inches). If you measure the end of your W-NIC antenna housing, you will see that it is too small to accommodate this 2.5-inch length.

To accommodate small packages, such as those on the W-NIC card, designers must use smaller basic antennas, or ones that are folded in some way. Two examples of this are the $\frac{1}{4}$ wavelength style, called a *quarter-wave antenna*, and the *folded dipole*. Quarter-wave antennas can be found as short vertical whips for cellular and two-way radios in cars. The flat, translucent FM wire antenna that came rolled up with your stereo receiver is an example of a folded dipole. A quarter wavelength antenna at 2.4 GHz is approximately 31.25 mm (1.25 inches) in length, which is still a bit long for the W-NIC case. Plus, a quarter wave antenna is highly directional when placed on its side. And without the so-called *ground plane* of a car (the metallic roof or trunk lid), they don't work very well. However, they can be attached to the outside of the W-NIC case and turned up vertically. In this case, they are more omnidirectional, covering the horizontal plane around them.

Antennas also can be shortened with loading coils. The most common example of this is the "pig's tail" antenna used for vehicle-mounted cellular radios. We don't see them much anymore because we have tiny hand-held cell phones. Those tiny cell phones illustrate another way to shorten an antenna.

An antenna's physical length may be shortened by using an appropriate length of wire wound around a rod-shaped form. Examples of this type of antenna are the shortened stubs that stick out from many cell phones and most cordless telephones. Cell phones operate at ~850 MHz (also 1800 MHz) and the newer cordless handsets at ~900 MHz, for which quarter-wave antennas would be about 8.3 cm (3.25 inches). The shortened cell phone antennas are often about 1 inch long, and some are even embedded in the cell phone case.

There is also a form of printed-circuit antenna that is quite small relative to a quarter wavelength. This is the style used with most embedded-antenna W-NIC cards. The antenna is formed from two copper traces, each in the form of a letter C. The somewhat circular shape of these antennas give them a more omnidirectional coverage than would be available from an imbedded horizontal antenna.

All of these shortened antennas are less efficient than their full-size cousins. That means that they do not transmit or receive quite as well. That factor may a lot of the difference in a marginal wireless link. Some of the manufacturers provide a means to connect a more efficient external antenna. If your W-NIC has a connector for an external antenna, you can improve the performance of your wireless network connection over a marginal path, thereby increasing the range and the maximum speed.

Performance

You will be able to determine the proper options for your W-NICs based on the information you learn in this book. Each installation of a wireless network is different, and, for best results, you must look at your particular environment. Factors to consider include the construction materials of the building or home, the distance from the AP, and potential sources of interference. The first step in this process is to make an educated guess of where you want to place the AP in relation to the wireless stations (PCs or laptops with W-NICs) based on the typical operating ranges and your environment. The next step is to test using actual WLAN components.

Performance testing is actually fairly easy with WLAN technology. Because the signal strength and unwanted noise vary so much with each and every wireless connection, the W-NIC manufacturers provide some pretty handy tools for measuring your link. You should definitely do some preliminary testing before you permanently mount the AP. Sometimes a minor adjustment can make a lot of difference. More on this later.

If you want to install wireless networking in a desktop computer, you will need a WLAN interface card that installs in an expansion slot. You may purchase a card that adapts your PC-style W-NIC to the appropriate bus slot in your computer. The most popular expansion slot is the PCI bus standard. However, a few manufacturers still offer the older ISA bus adapter cards. You can also get W-NICs that connect via the USB port. This is a really handy and noninvasive way to add wireless capability to a desktop computer, but it is cumbersome to use with a laptop, if you want the laptop to be truly portable.

Even though WiFi compatibility testing assures us that all these wireless products from various manufacturers can work together, you still might want to take the conservative approach. Make sure that a wireless PCI card or an appropriate adapter is available from the same manufacturer. Of course, there is less chance of such incompatibilities with a simple adapter than with an all-in-one PCI card.

Ad Hoc Network Operation

W-NICs can actually connect directly to each other, without an access point. This mode of operation, called *ad hoc*, allows two or more W-NICs to form an unsupervised wireless network. This can work quite well for very simple situations, such as one or two laptops in a conference room. It is particularly handy for transferring data files back and forth between the two machines. However, some of the advanced features available in a supervised network would be missing.

For example, most PCs with W-NICs get their Internet protocol (IP) addresses from the AP (through dynamic host configuration protocol [DHCP]). Along with the automatic IP assignment, they may get domain name service (DNS), gateway, and Windows Internet naming (WIN) assignments (see Chapter 3). Without an AP or a DHCP server on the bridged network, there is no way for the W-NIC to get its IP assignment. You would have to manually assign the IP address and the other parameters. This is a cumbersome process that would be beyond the average user's abilities and would require the intervention of a network administrator.

In addition, the ad hoc network may not be able to resolve hidden-station problems. When stations cannot see each other in a wireless network, they may attempt to transmit at the same time, causing unwanted interference to a third station. An AP helps mitigate this problem. The AP is also typically located in a central position, thereby potentially doubling the range over which stations may talk. Some APs may be equipped with higher gain external antennas, which also help extend the wireless range.

Wireless Access Points

An access point (AP) is a bridge or router that connects between the WLAN and wired network resources. These wired resources may be locally LAN-connected computers and printers, or they may be connections to the Internet, such as through a cable modem or digital subscriber line (DSL) modem. Essentially, the AP functions as a portal between the 802.11 wireless medium and other non-802.11 devices. A typical AP is shown in Figure 2.5.

Figure 2.5
A typical access point (AP) for SOHO use.

All APs perform essentially this same function. However, all APs are not created equal. The AP shown in Figure 2.5 is intended for "small office/home office" (SOHO) use. It has a fixed, internally mounted antenna, with a built-in WLAN transceiver and an Ethernet port. APs intended for commercial installations, such as larger buildings or campuses, will have provisions for an external antenna that is connected by means of a coaxial cable of an appropriate length. Commercial APs may also have provisions for more than one WLAN transceiver and more than one Ethernet port. There's more on commercial installations in Chapter 8.

Let's take a look at how a simple WLAN is formed. To form this basic wireless network, all we need is one AP and one or more PCs or laptops with W-NICs installed (Figure 2.6). In this figure, the wireless stations (PCs or laptops) connect wirelessly to the AP, and the extent or coverage area of the AP is indicated by the circle. The nonwireless devices appear to be outside the coverage circle, but that is only for clarity in the drawing; it really doesn't matter whether nonwireless equipment is within the coverage area. Also, for simplicity, we showed the AP's only local resource connection as a router that connects to the Internet.[5] This Internet connection is really not important in discussing the WLAN, itself, but it is the most common reason for creating a small home or office WLAN, so we include it.

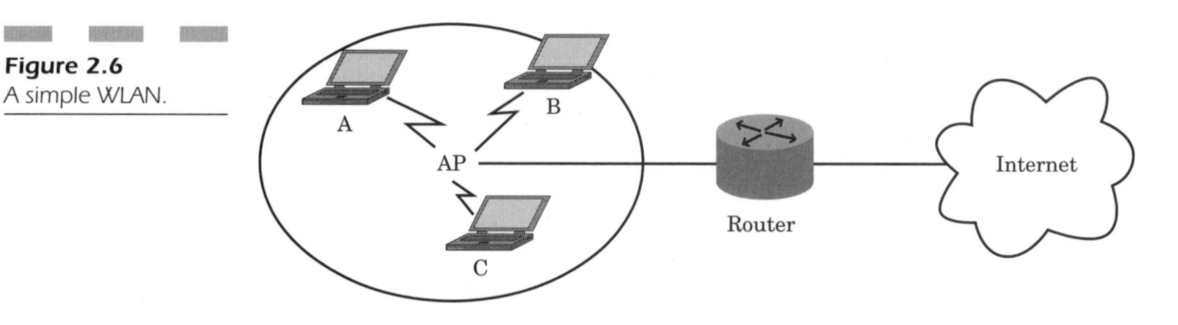

Figure 2.6
A simple WLAN.

WLAN Channels

In an 802.11b WLAN operating at 2.4 GHz, there are 11 frequency channels to choose from in the United States and Europe, with various other channel combinations in other countries. The channel center frequencies are in 5-MHz increments, starting at 2.412 GHz, and are numbered, conveniently, 1 through 11 for operation in the United States (13 channels are used in much of Europe, whereas only 4 are available in France and 1 in Japan). The 802.11b uses direct-sequence spread sprectrum (DSSS) operation. In addition to DSSS, there is another method that is used primarily in older systems.[6] You can look in the Appendix for complete information on these channels. All the technical details are not important now; you just need to know that, of the 11 available channels,

[5]Actually, this could be any type of wide-area data communications device, such as a cable modem, a DSL modem, an ISDN bridge/router, or even an analog modem. For that matter, it could be a wireless link to an Internet service provider. Except for the analog modem, all of these devices would operate as a simple router.

each AP is set to a one. Let's assume for this example that our access point is set to Channel 3.

When we start up a wireless workstation, let's say our laptop, with a properly installed W-NIC, will automatically begin to look for an AP. Unless we have set our W-NIC to operate on one particular frequency channel, it will scan through the available channels looking for our AP. Once it finds the AP (assuming it has an acceptable signal level), it will attempt to connect.

Tech Tip: Spread Spectrum in 802.11 Frequency Bands

In recent years, an innovative modulation method called *spread spectrum* (SS) has come into wide use in radio communications. This technology was originally developed by the military as a way to overcome radar jamming. The frequency of the radar essentially hopped from frequency to frequency to avoid jammers. Because the transmitter and receiver on the radar unit hopped to the same frequency, the return signal could always be picked up and the jamming avoided.

The communications uses of such a technology are obvious. Because noise and interfering signals have the same effect as intentional jamming, spread spectrum transmission can avoid these communication busters. Several types of spread spectrum operation are available. The easiest to implement is frequency-hopping spread spectrum (FHSS), which uses a set sequence of hop frequencies. In the 802.11 standard, FHSS can use 23 hopping patterns among 79 available channels in the 2.4-GHz band. Theoretically, multiple networks can share the same frequencies, just not at the same time. Any time a network has hopped to a particular frequency, the others become available for other devices.

Another method is direct-sequence spread spectrum (DSSS) and it can operate to 11 Mbps in the 2.4-GHz band (2.400–2.4835 GHz, a total of just over 80 MHz). Because it occupies a larger bandwidth,

[6]An older standard, permitted under the original IEEE 802.11 standard, uses frequency-hopping spread spectrum (FHSS). In North America and Europe, the FHSS band is divided into 79 channels, beginning at 2.401 GHz. The channels are numbered 2 to 80 and are 1 MHz apart. FHSS is limited to a data rate of 1 MBps and is no longer in wide use. Operation at 5.2 GHz is described in the Tech Tip on Spread Spectrum in 802.11 Frequency Bands.

DSSS channels are spaced 5 MHz apart, from 2412 to 2462 MHz, for a total of 11 channels in North America. Most of our attention in this book concerns DSSS operation in the 2.4 GHz ISM band.

The other band permitted by IEEE 802.11a begins at roughly 5.2 GHz. This band consists of three band segments at 5.150–5.250 GHz, 5.250–5.350 GHz, and 5.725–5.825 GHz. These segments are collectively considered part of the *unlicensed national information infrastructure* (U-NII) in the United States, and total 200 MHz between the three band segments. The exact channelization is shown in the Appendix. What is important is that new high-bandwidth WLAN services will occupy these 5.2-, 5.3-, and 5.8-GHz bands. IEEE 802.11a requires data operation at 6, 12, and 24 MBps and allows optional operation at 9, 18, 36, 48, and 54 Mbps. So the data rates higher than those allowed at 2.4 GHz will be in this band.

There is yet another unlicensed ISM band at 24 GHz, but it is only a 250-MHz segment, that does not have an inherent advantage over the lower 5-GHz bands.

Network Names

In standard 802.11 networking, each wireless network is assigned a network name. This is a little like the workgroup name in Windows Networking. This technique somewhat restricts the use of a particular network to devices that know that network name. For example, in a large building or a plant, you could use a confidential network name for the engineering department and keep the folks in accounting out. They would not know the network name for the engineering network and would be unable to infringe on the other wireless network. Actually, that might not be a bad thing. Of course, it would be more likely that the engineers would get into the accounting WLAN than the other way around.

The network name may be set to "any," which would allow the W-NIC to acquire service from any AP that does not limit network names. The downside to this feature is that it may be possible to detect the workgroup name and garner unauthorized data from a WLAN or even make an unauthorized entry into a WLAN. Fortunately, there are some safeguards you can take to limit this security risk. See Chapter 6 for a discussion of wireless security.

Connecting Wireless Networks

An AP is a station on the WLAN, just as each computer with a W-NIC is a station on the WLAN. In addition, the AP is a station (commonly called a *node*) on the wired Ethernet. One of the primary purposes of the AP is to bridge the wireless and the wired networks. However, nothing limits a wired network from having more than one WLAN attached. An idea of how this works is shown in Figure 2.7.

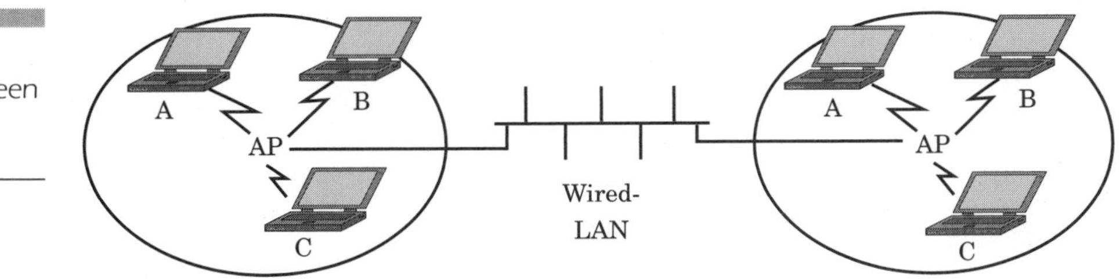

Figure 2.7 Connecting between WLANs with a wired LAN.

The AP can connect between diverse WLANs through an external network. This is one way in which wireless networks in more than one area can function together. For example, if you have a rather large area to cover, such as an office building or a campus (or a really big house), you will have to use multiple APs. Each will service any WLAN station within its area of coverage and be interconnected to reach other stations in the larger network or other resources such as the Internet.

At present, there are some technical issues in transitioning between two WLANs. Ideally, you would want to be able to walk from room to room, or from building to building, and stay logged onto the underlying network, despite going between two different WLANs. However, not all APs communicate with each other to allow this to happen. In some larger network environments, the wireless device is actually assigned a network address through a sophisticated authorization process. You would naturally want your authentication to map across adjacent WLANs, so you did not have to reboot your laptop. Special software is available to facilitate this.

In general, inter-AP communication is available only with the more complex (and expensive) APs intended for commercial use. It may not be available at all for the simple SOHO AP. Check with your AP manufacturer to see if it has implemented these features.

You could also have WLANs in two different cities, interconnected through a wired LAN and a wide-area link, such as a dedicated data line or even a VPN over the Internet.

Another variation of this scheme is shown in Figure 2.8. Here, two APs with special remote-access software (R-AP) connect the two wired LANs. Plus, you could have another WLAN at each end, connected to the wired LANs there. This would be a great scheme to get wired and wireless connections to a remote building. For that matter, you could use it to get between your mansion on the hill and the boathouse on the water. Well, you get the idea.

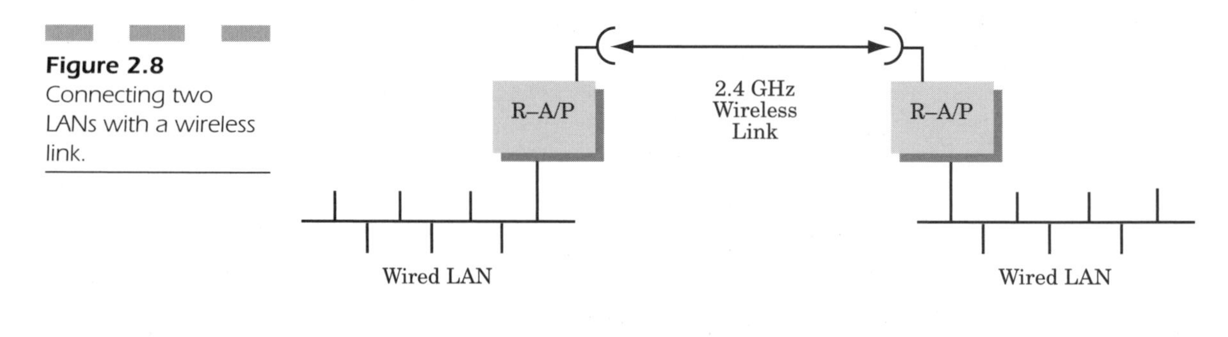

Figure 2.8 Connecting two LANs with a wireless link.

Thinking about Wireless Networking

Now it is time for a little "imagineering." You are going to learn an intuitive approach to WLAN operation. This will enable you to think through 90 percent of your WLAN installation issues. You will be able to "see" radio waves, or at least imagine that you can!

The best way to approach thinking about wireless radio transmission is to use some helpful analogies. Please note that "the map is not the territory." In other words, these analogies are useful for representing WLAN behavior, but they only approximate what you would experience with real, live radio waves.

Radio propagation is a wonderful mystery. Most of us go through life turning on the TV or the car radio without understanding a thing about how it all works. However, if you have ever hooked up your TV to a crude antenna, such as "rabbit ears" or a UHF loop antenna, you have an idea how radio waves work. Move the simple antenna this way or that and the signal will go haywire. When you position the antenna to get a better signal, you are experiencing some of the effects of *antenna alignment*, *directivity*, *polarization*, and *signal capture*. The better arranged the antenna is, the better the signal will be. There are different styles and polarizations of antennas, and they can be pointed toward or away from the transmitted signal. Antennas naturally respond in some directions better than in others, a property called *antenna pattern*. We discuss antennas later in this chapter.

Now, walk around in front of the antenna. Watch the TV signal change. You may find that touching the rabbit ears makes the signal stronger or weaker. You may notice the influence of other objects, as your body may be a *reflector* or *absorber* of the transmitted TV waves under certain conditions. It is important to take *reflections* and *attenuation* into account when setting up your wireless network.

If you try to get TV reception inside a large building or use your car radio in an underground parking garage, you will experience *signal fading*. The same thing happens when you drive beyond the range of your favorite FM station. In one case, the fading is due to the absorption of the signal; in the other case, fading is due to the progressive weakening of the signal over longer distances. We call this phenomenon *attenuation* or *path loss*, regardless of the cause.

With television signals, to reduce such effects, we can use external antennas, coaxial cables, and amplifiers to strengthen and distribute the TV signals. We do the same with all radio signals, and a WLAN is just a specialized RF transmission. We will use some variations of the same tricks to enhance performance of our WLAN.

One of the best ways to understand how radio waves are conducted and propagate is to use some familiar objects as analogies. Although this method is imperfect, it will give us some of the intuitive understanding we need to plan and implement a WLAN.

We have said that WLANs operate at frequencies that are basically *line of sight*. In reality, this is the "sight" of the radio waves, which is different from visible sight, but it works pretty much the same way. Figure 2.9 illustrates the line-of-sight nature of an ordinary light bulb.

As you can see from Figure 2.9, light is transmitted in different ways through different objects. It can be partly blocked, or attenuated, by semitransparent objects such as darkened glass or shade screens, reflected off surfaces, or blocked entirely by solid objects. In the real world, the light meter behind the opaque object probably would get some reflected light off walls or other objects, although the direct rays would be blocked.

This is much like what happens with the radio waves of your WLAN. The main difference is that we have to imagine (or measure) what happens to the radio waves, because we cannot see them. We also have to

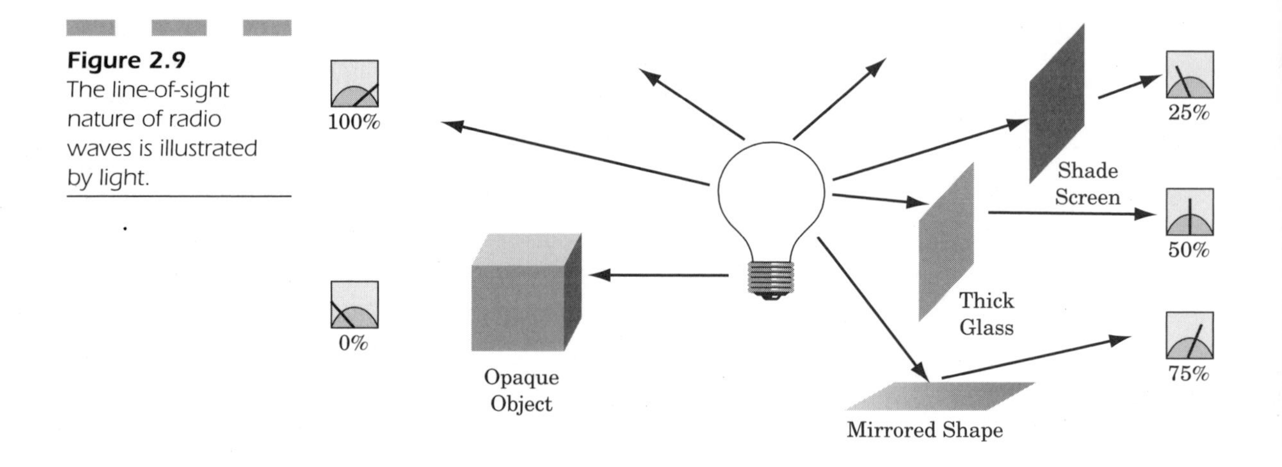

Figure 2.9
The line-of-sight nature of radio waves is illustrated by light.

anticipate the effect of objects in our wireless environment, based on how those objects affect those radio waves.

Radio waves are affected primarily by the thickness (density) or conductivity of objects. For example, the attenuation of a radio wave is greater through a brick or stone wall than through a gypsum wall board or a wood wall. Conductive objects such as air ducts, metal doors, filing cabinets, and even reflective coatings on glass can totally block a WLAN signal, in some cases. Other times, conductive objects reflect the WLAN signal in the same way as a mirror as shown in Figure 2.9.

In other cases, the receiving WLAN card might receive some direct signal, and some reflected signal—a real problem for radio signals. This combined signal causes *multipath* interference and may decrease the range or useable data rate in a WLAN signal or make the signal totally unreadable. For that reason, you will see that some W-NICs offer "superior multipath performance" as one of their features.

If we ignore color in the light analogy, you can get an idea of reflectivity. It is easy to see that we get more reflected light from white objects than from black ones. Just look around the room and you will see that a whole range of objects fall in between, from a brown rug or desk to the beige or gray shades of cabinets, office furniture, or PC. For the most part, you are always gazing at reflected light. Now glance at the light fixture itself. It is very bright in relation to the reflected light from the objects in the room. Now look at the same light through a mirror, and you will see that it is almost as bright as the original, particularly if the mirror is close to your eyes. You have just illustrated the reflectivity phenomenon of WLAN radio waves, except that they depend on *conductance*, a different property of objects, for the signal reflection.

Tech Tip: Light Signals

OK, here is another mind game you can play. We usually think of light as just bright or dim in these radio analogies. However, we can use light to convey information. For example, we regularly use traffic signals to control intersections. You may know that the human eye's ability to see color is very sensitive to the light intensity. At reasonable intensities, we see all the colors, but as the intensity drops, we are less able to distinguish color. In near darkness, we see in black and white because we need the higher light intensities to activate the color receptors, called *cones*, whereas the more sensitive noncolor receptors, called *rods*, can still detect usable light.

In a way, this color versus black-and-white behavior of our eyesight mimics the behavior of WLAN signals. The higher data rates, above 2 Mbps, require increasingly more signal to function, much like our color vision. The electronic brain of the WLAN card compensates by conveying information more slowly, so that our network still functions, albeit not at full capacity.

For any given signal source, whether radio or light, we can distinguish (receive) based on the intensity (power or directivity) of the source and the distance from the source. So, we can easily see the red traffic light close up, but it becomes increasingly difficult to distinguish as we get farther away. Fortunately, red was chosen for this purpose, because it is one of the last colors to fade from recognition. But the intensity of the source is also important, which is the reason we can easily see that Mars is red, even though it is millions of miles away. Perhaps it is signaling something!

Another very useful light analogy is found in devices that direct the light beam. A fluorescent light fixture directs more of the light into the room. A spotlight sends more light in one direction than in others. This is very similar to the effects of directional antennas on radio waves. In Figure 2.10, for example, a fluorescent light fixture sends light mostly in one direction, in the same way as a directional radio antenna. Parabolic reflectors may be used effectively to concentrate both light and radio waves into relatively narrow beams.

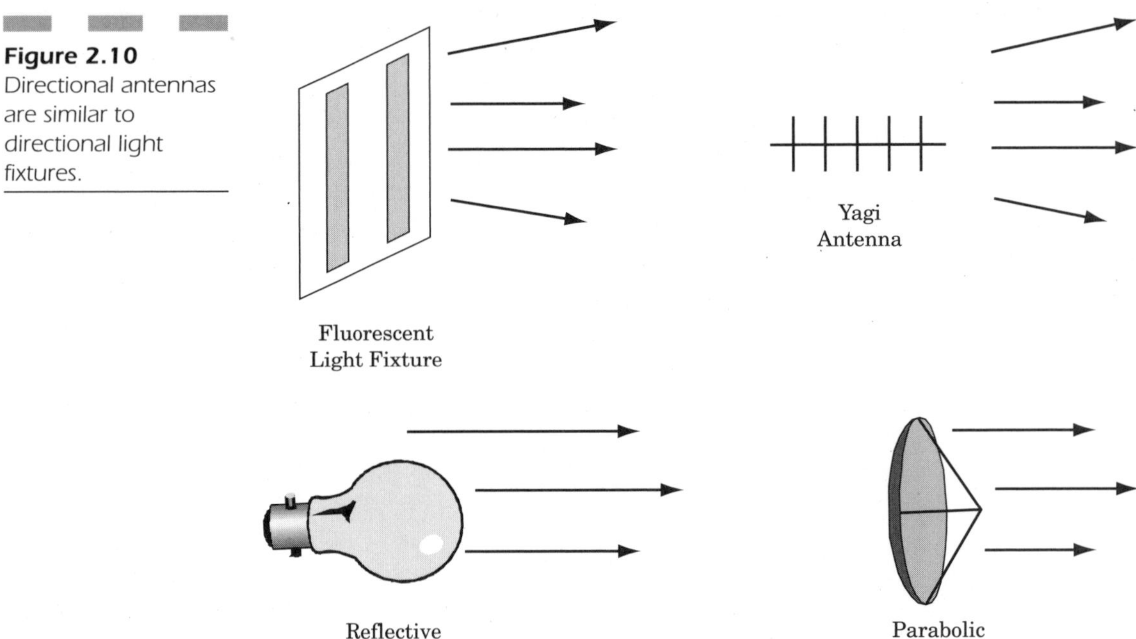

Figure 2.10 Directional antennas are similar to directional light fixtures.

We have lots of other antenna types in WLANs, as you can see from the section on antennas later in this chapter. However, many of the principles that make these antennas effective can be understood intuitively from your existing knowledge about the behavior of light.

Wireless "Plumbing"

Another useful analogy is the water pipe or water hose. We use pipes in our homes to distribute water to where it is needed. You can think of a water hose as being somewhat like the coaxial cables we use to connect antennas to our WLAN equipment. Both have two ends, both have connectors, and both transport something from one end to the other. A large-diameter hose can move more water; a large-capacity coax cable can handle more RF power. A long, narrow hose will deliver lower water pressure and, consequently, less water; a long coax with poor transmission characteristics (lots of loss) will deliver less RF signal. Kinking the hose will partly block the flow of water; kinking the coax will cause a signal loss. Now, so that there will be no misunderstanding, there are some major differences, but the similarities are still very useful in thinking about RF cabling. As a matter of fact, at microwave frequencies, the wave guides that function as very low-loss coax replacements are informally called "plumbing!"

Oh, by the way, you can also pump up the water pressure to deliver more water through a lengthy or smaller pipe. Likewise, you can *amplify* the RF signal to compensate for losses in a lengthy or lossy coaxial cable. We're going to stop this water "think" now, before we all get soaked.

As with light, we can position the RF source to help us receive signals at greater distances or beyond structures that would block the signal. In the same way that there are highway lights on tall poles to cast light for longer distances, you will see cellular antennas mounted on towers or buildings to increase their coverage. With WLAN technology, we can add external antennas and position them so they carry the wireless signals above or around obstructions. We just have to be mindful of the fact that the cable length between a W-NIC or AP and the antenna causes a loss of signal strength. We will have a lot more on the siting of antennas and coax cable losses later.

Choosing a W-NIC

Choosing a W-NIC is fairly straightforward. You should first determine the format or type of card that your computer requires. There may be more than one choice open to you, depending on the type of computer you have. You must determine which choice is best for your intended use and the conditions of the installation. This may involve convenience factors, such as dealing with extra boxes and antennas, or it may be somewhat dictated by the need for an external antenna to get the coverage you want. Let's take a look at the options for W-NICs.

Laptop W-NICs

There are differences between W-NICs, even though all W-NICs are basically designed to interoperate. You will want to make certain that the W-NICs you buy are compatible with your hardware and software environment. The most common form of W-NIC is the PC (or PCMCIA) card. Most of these require what is called a "Type II" slot in your laptop or desktop computer. Almost all recently manufactured (within the past 5 years) laptop computers will accept a Type II card. However, some PC cards effectively take up two slots because they offer built-in connectors for telephone or LAN connections. If you have one of these cards, you will have to remove it[7] before your W-NIC can be inserted. If this is a real bother, you could just consider converting everything to wireless, so you won't need the wired LAN card.

Desktop W-NICs

If you have a desktop computer, you will need to purchase an appropriate adapter card for the PCMCIA W-NIC. These adapter cards are available in PCI-bus and the older ISA-bus formats but not necessarily from all manufacturers. Some wireless manufacturers only offer PCI-bus adapters because that is the newer bus standard. Most desktop computers have open PCI slots, and some may have both PCI and ISA slots. The desktop computer case must be opened to install the adapter card—power off *and* power cord unplugged, please! You may have to install the adapter first and then plug in the PCMCIA card, depending on the adapter design. Some manufacturers offer a combined card with the W-NIC function included, although the adapter card approach is more common.

Keep in mind that most PCs are a big metal box. WLAN signals do not pass through metal, although they may manage to get some signal around the side or top of the computer. This is a good time to think of an external antenna, if your desktop is a good distance from the AP and oriented in the wrong direction. Think of the AP as a bright light and imagine whether your W-NIC antenna will be in the shadow or the light, as a result of the computer case and other metallic objects.

[7]You should only remove a PCMCIA card by first properly shutting the card down, if the computer is turned on. This can be accomplished through a utility program supplied by the PC manufacturer. If you do not know where to find this program, it generally is one of the applets in Start/Settings/Control Panel, accessible in the Start menu on the task bar in Windows 9x and later operating systems. In many cases, you can also safely remove a PCMCIA card if the laptop is off and is "shut down," not just hibernating. In most cases, the new card can be inserted into the slot while power is on, and the proper drivers will be loaded automatically.

USB W-NICs

The Universal Serial Bus (USB) WLAN adapter is another alternative to Type II PC cards and adapter cards. These clever devices make a connection to the computer via the USB port. So, you could actually use a W-USB (cool new term!) to connect a laptop or a desktop. There are some real advantages here. First, you can plug and unplug the USB adapters while the power is on. Second, you can move the W-USB around to maximize signal strength (USB cords may be as long as 10 feet or 3 m). Third, you can avoid taking your desktop computer apart and the constant plugging/unplugging of other PC cards in your laptop. Forth, you can easily upgrade to new wireless technologies, because you don't have to open your desktop. Fifth, you can move the W-USB easily to different computers or even trade among your friends.

However, with all neat features, there are some disadvantages. First, you have another little box attached to your laptop, which may be clumsy and a bother. Second, you have to have a laptop/desktop with a USB port and run an operating system that supports it. Third, USB-I can support only data transfer up to about 5 Mbps, which is theoretically below the 11 Mbps of 802.11b and well below the 24 to 54 Mbps capability of 802.11a. A newer USB-II standard can operate at nearly ten times the speed, but not all interfaces support it. Fourth, the W-USB adapters are normally more expensive than the simple PC versions. Fifth and finally, how are you ever going to hold the W-USB box on your lap along with your laptop, if you really do use a laptop on your lap? Well, it is not a perfect world, and you will have to choose the type of wireless adapter that is best for you.

Operating System Drivers

The next step in choosing a W-NIC is to determine operating system (OS) compatibility. This is important because each W-NIC must interface to the host OS through a *driver* program. Each driver is uniquely designed for a particular OS, such as Windows 98, 2000, or XP; Apple OS X, or Linux (in its various versions). The documentation or packaging for a W-NIC should show system requirements and compatibility. In general, you can probably assume that any OS in the same family can take the same driver. For example, Win 95 and Win 98 generally use compatible drivers, as do those OSs in the NT family, Win NT, Win 2000, and Win XP.

W-NIC operation requires a driver and a setup/control program, but both of these are relatively small, so you should not have disk storage problems when you load these programs. Just be sure to keep the setup disk available, in case you ever have to reinstall the W-NIC.

Compatibility and Standards

There are several wireless networking standards in existence and a number of proprietary systems. We recommend that you pick an international standard, such as IEEE 802.11a or .11b, to get the most compatibility. It is the most widely used WLAN system and will ensure that you can use your wireless components in a variety of situations for a number of years. The 802.11b cards operate at 2.4 GHz (2400 MHz), and the 802.11a cards operate at 5.2 to 5.8 GHz (often called 5 GHz). The 2.4-GHz cards are currently the most popular, but the 5-GHz band offers higher speed and less interference from competing wireless technologies. Many of the 5-GHz cards and APs offer backward compatibility to 2.4 GHz, so that they can operate in both bands. Look for the frequency bands and standards on the packaging and in the documentation for WLAN components.

Antennas and Cables

Our final topic in this chapter on wireless basics is the wireless antenna and its necessary cabling. Even though it is not visible in most W-NIC cards, an antenna is there, under the plastic cover. Most of the APs designed for SOHO use also have hidden antennas beneath the plastic case. A few manufacturers have designed their cards and APs with one or two tiny antennas that can be positioned for best results. Other manufacturers of cards and APs allow you to attach an external antenna with a short cable. You may be able to obtain a variety of external antennas with different characteristics. Coaxial cables may be used to position the antennas some distance from the W-NIC or AP.

Both antennas and cables are totally passive devices requiring no power. We will give you a good basic understanding of these simple antennas and the effects of adding various lengths of cables to your antenna installation, if you choose to use an external antenna.

Antenna Characteristics

Antennas are devices that convert electrical signals into electromagnetic fields that can propagate through space. Likewise, they can capture electromagnetic fields and convert them back into electrical signals. We transmit a signal into the air (commonly considered *free space*) by coupling the signal to an appropriate antenna. Conversely, we couple an antenna to appropriate circuitry to receive (or recover) the information in an electromagnetic wave.

The operation of antennas is a tiny bit technical, so there are two paths in this section. We explain antenna operation and cable losses in general terms in the text of this section. If you want additional technical information, we refer you to a series of Tech Tips that provide additional details and explanations.

Electromagnetic (EM) fields, or radio waves, have two fundamental properties, *frequency* and *wavelength*. The frequency is the rate of change of the EM field as the wave travels through space; this is identical to the frequency of the corresponding electrical signal. The wavelength is the distance the EM field travels during one cycle of the EM field. See the Frequency versus Wavelength Tech Tip for the details of this process.

Tech Tip: Frequency versus Wavelength

The wavelength of an electromagnetic (EM) field is the distance the wave travels to complete a single cycle of amplitude. For each frequency, there is a corresponding wavelength. The two are directly related by the speed of light in so-called free space—a perfect vacuum. In a less than perfect medium, such as a copper wire, a derating factor, called the *propagation factor*, provides the correct wavelength value. The wavelength is normally given the symbol lambda (λ), frequency the symbol f, and the speed of light the symbol c. The wavelength formula in free space (propagation factor of 1) is:

$$\lambda = c/f$$

In metric terms, the speed of light in free space is a constant of approximately 3×10^8 m/sec. If we pick some convenient units for our 2.4-GHz frequency band, we can reduce the formula to:

$$\lambda(\text{cm}) = \frac{30}{f\,(\text{GHz})} = \frac{30}{2.4} = 12.5 \text{ cm or} \sim 5 \text{ inches}$$

Thus, the wavelength at 2.4 GHz is approximately 12.5 cm, or about 5 inches. Consequently, a half wave would be about 6.25 cm (2.5 inches), and a quarter wavelength would be about 31.25 mm (1.25 inches).

The wavelength of a WLAN transmission influences the size of the antennas we can place within our W-NICs and APs. The effectiveness of the built-in antennas in W-NICs is compromised by their small size, necessitated by the size of the PC card. Bottom line: these antennas are not very efficient. As a result, you may want to use external antennas on the APs, or on the W-NICs, if possible, to increase the signal strength to and from your portable computing devices. See the Tech Tip on Antenna Size for more information.

Antenna Pattern

Most antennas receive and transmit signals better in some directions than in others. This is called the *antenna pattern*. In general, the patterns of antennas decrease the signal in certain directions to increase it in others. For example, it is more useful to increase the signal from side to side, around a W-NIC or AP, than straight up or straight down. Fortunately, simple antennas do this naturally, and we adjust length, shape, and reflective elements to help them along.

The small antennas on our W-NICs are set up to receive signals pretty equally in any horizontal direction, despite a little blocking that comes from the laptop itself. Some W-NICs have small antennas that can be turned upward from the card for better signal reception. Other W-NICs have a connector for an external antenna. External antennas may be necessary to cover wireless devices that are beyond the normal range of coverage of your AP.

Antenna Polarization

Antennas also possess a quality called *antenna polarization*. Polarization refers to the orientation of the electrical (and magnetic) fields as the radio wave passes through space. We will not get into all the details here, but you should understand that antennas can be vertically or horizontally polarized and that cross-polarization can greatly affect signal strength. In practice, this effect can cause signal loss as high as 10 to 1 when transmitting and receiving antennas are at different polarizations. See the Tech Tip on Antenna Pattern and Polarization at the end of this chapter for more information on these two subjects.

All antennas, even the built-in kind, have pattern and polarization characteristics. You can sometimes improve reception by moving the W-NIC antenna with respect to the AP or by reorienting the AP to achieve better coverage in a particular area. With a laptop or PDA, this may mean that you move it slightly by turning to the side. If polarization is the problem, the type of W-NICs with the tiny rabbit-ear antennas can be reoriented to help out. This is best done while looking at the signal strength indicator in the link-test tool supplied with most W-NICs. In addition, you may want to add an external antenna, if your W-NIC is equipped for this.

External Antennas

SOHO APs normally do not provide for an external antenna. They provide built-in antennas that have some performance gain over the small antennas in W-NICs. APs intended for commercial use normally have a provision for a wide variety of external antenna options. There are a few APs intended for SOHO use that allow you to connect and add-on antennas.

External antennas are available in a variety of styles. Table 2.1 lists many of these antenna types and their characteristics.

External antennas are used for the following reasons: 1) to increase range, 2) to provide coverage where a built-in antenna would be blocked, and 3) to provide coverage in a certain direction (perhaps to decrease interference). Each antenna in Table 2.1 may be used to enhance the communications link, based on one or more of these factors. For example, an AP may need a panel antenna to cover a wide area outside a building, where the wide "fan" pattern would be perfect. Or a stacked dipole antenna, such as the desktop-mounted range extender shown in Figure 2.11, could be used to regain coverage around a blocking object, such as a file cabinet or a concrete and steel building structure. The Yagi and parabolic antennas are particularly useful for greatly extending the useful link distance of wireless networks, along with the ability to reject interfering signals from other directions. Any antenna that produces a relative signal gain will also allow the operating speeds to be maximized. Remember that 802.11 W-NICs adjust the data speed downward to compensate for poor signal strength, so increasing the signal strength by using a higher gain antenna will allow the maximum data rate to be greater.

TABLE 2.1
Antenna Types and Characteristics

Type	Mount/Use	Pattern*	Polarization	Gain (dB)	Relative Cost
Quarter wave whip	Mobile	Omnidirectional	Vertical	3	$
Stacked dipole	Desktop or wall/pole	Omnidirectional	Vertical	6	$$
Panel	Wall	Unidirectional (150° fan)	Vertical	10	$$$
Yagi/Uda	Pole	Unidirectional (15° beam)	Horizontal or vertical	12–15	$$$
Parabolic	Pole	Unidirectional (5° beam)	Horizontal or vertical	18–25	$$$$

**Approximate 3-dB beam width in degrees. Omnidirectional antennas cover 360° in the orthogonal plane (horizontal plane for a vertical polarization).*

Figure 2.11 Range extender.

Coaxial Cables

An external antenna is connected to the W-NIC or AP with a coaxial cable. Some antennas come with a short length of cable, as shown in Figure 2.11. Coaxial cables come in a variety of types and lengths, and several connector styles are available. It is important to understand a little about these cables so that we can understand how we can best use the proper type and length of cable with our wireless networks.

The coaxial cable is a fundamental component of most RF transmitting systems. That is to say, most RF transmitting and receiving devices use some form of coaxial cable between each device and its antenna. Some examples are the coaxial cables we use to connect our televisions, whether we use an external antenna or simply connect them through the cable TV provider's distribution system. Another example is the antenna cable that connects our car radio to the antenna on the fender (or perhaps internal to the windshield). For our purposes, we will discuss only the coaxial components that are used in WLANs.

Coaxial cables come in different sizes and with different values of impedance. Generally, the larger diameters of coaxial cable have less signal loss. All coaxial cables progressively lose signal or attenuate the signal. The *attenuation* is proportional to the length of the cable, so for any particular type of coax, the longer the cable, the greater the loss. The same is generally true of cable diameter, although special types of dielectric (the insulating layer between the center conductor and a shield) are available with much lower losses.

Coaxial cables for WLAN use are varieties with a 50-Ω impedance. Most coaxial cables for communications use 50-Ω cable, unlike the popular 75-Ω cable used in television reception. The most popular, and most available, 50-Ω coaxial types are RG-8/U, RG-58/U, and RG-178/U, in sizes from largest to smallest. RG-8U is slightly smaller than 1/2 inch in diameter, RG-58U is a bit smaller than $\frac{1}{4}$ inch, and RG-178U is smaller than $\frac{1}{8}$ inch. Metric conversions are left to the student. In the multiple-Gigahertz world, a number of ultra-low-loss cable types are available, but they are dimensionally similar to these classic RG cable types. A list of common coax types and their characteristics is given in Table 2.2.

There are also versions of these cables that are intended for so-called plenum spaces, which are the air ducts usually found in commercial buildings. Plenum-rated cables are generally somewhat smaller in diameter than nonplenum cables. You will probably use nonplenum cables in most home and outside locations, whereas plenum-rated cables may be required in the above-ceiling grid areas of commercial buildings. Cable color is usually black, especially for cables that are UV-stabilized or intended for outside exposure to sunlight. However, other colors, such as white or gray, are available and may be more suitable for indoor locations.

TABLE 2.2
Coaxial Cable Characteristics

Coax Type	Impedance (nominal)	Loss*/ 100 feet @2.4 GHz	Loss*/ 100 feet @5.2 GHz	Typical OD** inch (mm)	Relative Cost
RG-8/U	50Ω	16 dB	25 dB	0.403 (10.2)	$$
RG-58/U	50Ω	38 dB	62 dB	0.192 (4.9)	$
RG-178/U	50Ω	35 dB	56 dB	0.075 (1.9)	$
Semirigid helical dielectric	50Ω	2.6 dB	5.0 dB	0.875 (22.3)	$$$$

**Loss shown is typical for cable type. See manufacturer's specifications for your cable.*
***OD, outside diameter, shown for nonplenum types.*

Coaxial loss can effectively negate any antenna gain. One of the primary reasons you might need to use a high-gain antenna is to compensate for cable loss. For example, with a relatively short length of cable, you might determine that a 6-dB antenna is needed for coverage from the roof of a structure. However, if you must locate the AP farther from the antenna, such as higher on an antenna pole, the additional coax might increase your loss by an additional 6 dB. To compensate, you will need to use a 12-dB gain antenna or perhaps increase your signal by 6 dB with an amplifier.

By the same reasoning, you should try to minimize coaxial cable loss. That means that cable length should be minimized and that low-loss coax types should be used whenever possible. We cover some of these issues and how to plan out the allowable loss in particular situations later in the book. For the time being, just remember that you want to minimize signal losses caused by coaxial cables.

Coaxial Cable Connectors

The three most common connectors in Gigahertz-class systems are the N-type, the TNC, and the SMA. In addition, many manufacturers use semicustom designs to prevent the connection of unauthorized antennas and amplifiers. The federal rules for RF exposure are fairly strict and it might be possible to take a unit out of compliance by adding an antenna with too much gain. However, most add-on antennas use higher-loss coax cables because of cost, so the gain of the antenna is often compensated by the loss of the cable.

Figure 2.12 shows several of the more common connectors and some samples of coax cables. You can see that there is quite a variation, not only in size but in other characteristics. All of these connectors, the N-type, the TNC, the UHF mini, and the SMA, are considered low-loss constant-impedance connectors. For more information on coaxial cable, connectors, and connection methods, see *LAN Wiring*, by this author.

Figure 2.12 Connectors used with WLAN antennas (Type N, reverse TNC, UHF mini, and SMA).

You should consider the additional loss from connectors in your antenna connection. Each connector set can add up to 0.5 dB to the cable loss, as calculated from the cable's loss per foot, so you actually introduce extra losses in a cable that is made up of several smaller linked cables. Further, cables are normally terminated with a male-style connector, so a barrel (F-F) connector must be used to join the two ends. That is actually *two* connections, with up to 2×0.5 dB (or 1 dB) of insertion loss.

Loss/Gain Estimates—Decibels (dB)

Most of us have a little difficulty thinking in decibels (dB). This is really easier than it looks. When we use the decibel, every calculation is reduced to addition or subtraction, something everyone can master.

Decibels are just a logarithmic expression of the numeric gain or loss in a circuit. We measure gain and loss as a power ratio; that is, the ratio of output power to the input power. So an amplifier that takes a 5-mW input and increases that to 10 mW would have a numeric gain of 10/5 = 2. Similarly, a cable that attenuates an input power from 10 to 5 mW would have a numeric gain of 5/10 = ½, which we speak of as a 2:1 loss.

When we get larger power differences, the numbers get very large. For example, power output/power input =1000 mW / 1 mW = 1000 and, similarly, 4,000,000 μW/ 1 μW = 4,000,000. Because we are really concerned with the ratio in gain/loss calculation, it is easier to express this in logarithmic form. The decibel (one tenth of a bel) is defined as

$$10 \log P_o / P_i$$

where P_o is the output power and P_i is the input power of a circuit or device.

Therefore, the numeric power ratio of 2 is 10 log (2) = 3.010 dB, or approximately 3 dB. The large ratios of 1000 and 4,000,000 are now simply stated as 30 dB and 66 dB! Pretty cool, huh? Well, everybody needs a cheat sheet to understand this. You will notice that the ratio of 2 translates very close to 3 dB and a numeric gain of 4 is ~6 dB. A ratio of 10 is exactly 10 dB, 20 is 13 dB (or 10 dB + 3 dB ... twice 10 is 3 dB more), and so forth. There, you can compute decibels in your head, just like a real radio engineer. And you thought this was difficult! Table 2.3 lists the more common units.

One last item and we will move on. Power levels in RF work are almost always expressed in relative terms; that is, in decibels relative to some convenient power level. For example, in WLAN work, the output power levels of the W-NICs generally range from 10 to 20 mW. So we express those levels in decibel terms relative to 1 mW (1 milliwatt), or a 0 dBm power level (decibel referenced to 1 mW ... 10 log [1 mW/1 mW] = 0). So, 1 mW is 0 dBm, 10 mW is 10 dBm (remember a ratio of 10 is 10 dB), and 20 mW is 13 dBm. You will often see W-NIC cards rated at 10 dBm or 13 dBm and now you know the rest of the story...this 3 dB difference is double the power! Wow, that really is a difference!

By the way, antennas are said to have "gain" but they do not really amplify the signal. They just flatten or beam the power into fewer directions, thereby increasing the signal in a particular direction. So they have "effective" gain. If we multiply the applied power by effective gain, we get *effective radiated power* (ERP), sometimes called *equivalent effective radiated power* (EERP) for those who still don't get it. We will do more with output power, cable loss, antenna gain, and path loss in later sections.

TABLE 2.3

Handy Decibel (dB) Rules of Thumb

Power Ratio	Numeric Ratio	Decibels (dB)	Example	
1:1	1.0	0	Unity gain	0 dB
2:1	2.0	~3	Double or half	3 dB or –3 dB
3:1	3.0	~4.8		
4:1	4.0	~6	Lots of antennas	6-dB gain
10:1	10.0	10	Easiest to remember	
20:1	20.0	~13	Brain teaser—twice 10 dB is only 3 dB more	
100:1	100.0	20	Remember 100 times is only...	20 dB
400:1	400.0	~26		
1000:1	1000.0	30	And 1000 times is...	30 dB

**Loss is often shown as a positive number, although technically it is a negative gain. So a cable loss of 3 dB would be a –3 dB gain. You can calculate the resulting output power with simple addition and subtraction. For example, a 0-dBm signal plus a 6-dB gain would yield a 6-dBm output.*

Next we have a series of Tech Tips for those who want to know more. You can skim over them or skip them altogether if you don't want to concentrate on the technical details now.

Tech Tip: Antenna Pattern and Polarization

An ideal antenna would convert EM (radio) waves to and from electrical energy equally in all directions. Such a theoretical point source antenna is called an isotropic antenna. If you looked at the pattern from an isotropic antenna, it would look spherical (circular in all planes), like this:

Isotropic Radiator

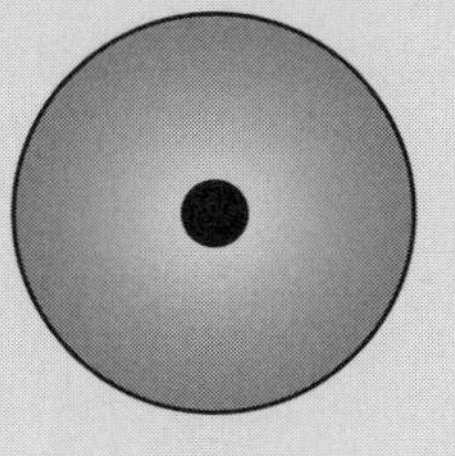

One of the problems that occurs with quarter and half wavelength antennas is that they have a moderately directional pattern. That means they receive radio signals much better in some directions. Specifically, quarter and half wavelength antennas receive signals better to the side than toward the ends.

Quarter Wave Antenna

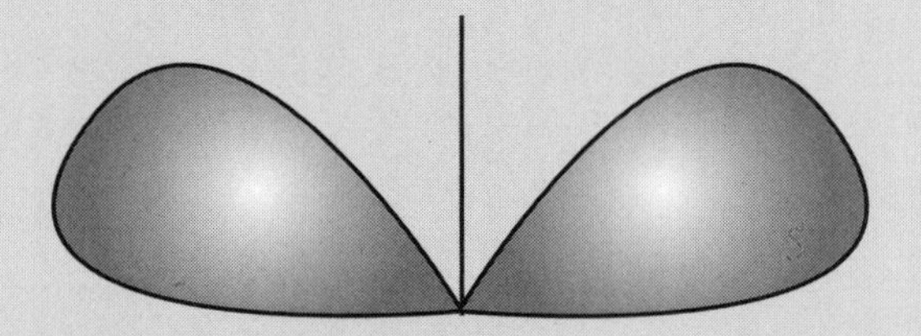

Side view of pattern

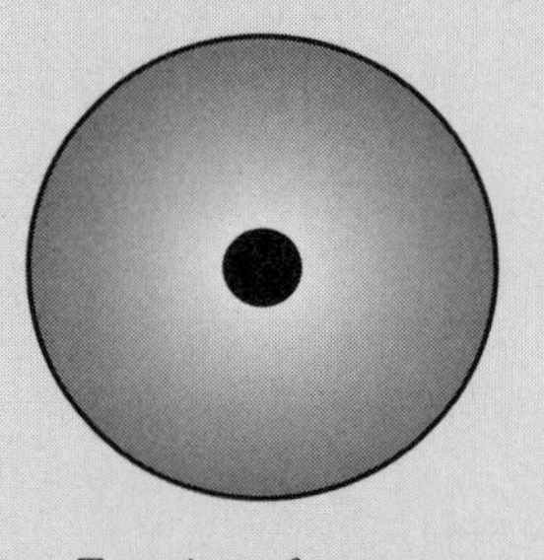

Top view of pattern

If you point a quarter wave antenna directly at a signal source, you'll decrease the strength of the receive signal; whereas, if you point the same antenna vertically, so that the source of the receive signal is to the side, you will increase the signal strength. Thus, alignment and pattern are very important in choosing the proper antennas for WLAN devices.

In addition to pattern, the signals from an antenna may be polarized, much in the same way that light is polarized. For example, a quarter or half wave antenna that is vertically mounted will transmit vertically polarized EM fields. Consequently, a corresponding receive antenna also should be vertically polarized to properly convert the vertically polarized transmission.

If there is a mismatch between transmission and reception antennas, a certain amount of signal loss occurs. For example, if the antennas on the AP are vertically polarized and the antennas on the W-NICs are horizontal, over 90 percent of the signal may be lost. To compensate for this effect in the WLAN antennas, with the relatively small W-NIC size, the antennas may be externally mounted, so that they are placed vertically and aligned with the AP antenna. Some W-NICs have small vertical antennas that are actually part of the card that extends outside the laptop. These small antennas may be rotated vertically, as needed. Other W-NICs have a tiny connector on the end of the card so you can connect an external antenna. Several types of external antennas are available, including full-size omnidirectional (equal signal in all directions, 360°, horizontally) and directional (maximum signal in one direction) antennas.

Built-in W-NIC antennas are often printed circuit board antennas that are folded or curved to compensate for directional and polarization effects and to fit the antenna within the confines of the W-NIC case. AP antennas have much more flexibility in size and design, simply because the AP case is much bigger.

Tech Tip: Antenna Gain

Antenna patterns may be used to advantage to create directional antennas that provide more signal strength than we would get from a nondirectional antenna. The theoretical point-source antenna, which transmits or receives EM energy equally in all possible directions, is used as our reference. We say that the theoretical isotropic antenna has a gain of 0 dB (a numeric gain of unity, or 1). With the isotropic antenna as a reference, any antenna that has more signal strength in any direction is said to have a gain proportional to that increased signal strength.

Virtually all practical antennas have some amount of gain relative to the isotropic antenna. The letter *i* is used to distinguish gain over an isotropic antenna. Thus, the quarter wave antenna is generally considered to have a gain of about 1.5 dBi, and a half wave antenna of about 2 dBi. If the desire is to produce higher omnidirectional gains, we can adjust the length to certain submultiple wavelengths or stack vertical elements to increase gain from 2.5 dBi to 6 dBi. The gain comes from limiting the signal from other directions, which generally flattens the antenna pattern compared with a quarter wavelength antenna.

We can increase the gain of an antenna by limiting its horizontal, or azimuth, coverage to less than 360°. Such an antenna is called directional. Two popular types of directional antennas are the Yagi-Uda array (commonly called Yagi) and the parabolic antenna (commonly called a dish). Yagis that are used at 2.4 GHz consist of an array of 12 to 15 closely spaced antenna elements that are enclosed in a plastic radome for protection from wind and rain. The array points in the direction of greatest signal gain and typically has a gain of about 12 dBi. Multiple Yagis may be put in an array to increase the gain even further.

A dish antenna, which is usually about 12 to 18 inches across, can produce 15 to 25 dBi of gain at WLAN frequencies. A parabolic antenna's gain is proportional to its size, within certain limits, so the gains change with size.

Here's a table that shows typical antenna gains:

Isotropic antenna	0 dBi
Quarter wave	1.5 dBi
Half wave	2 dBi
5/8 wave	4.5 dBi
Vertical array	6 dBi
Yagi-Uda array	12 dBi
Parabolic dish	15–25 dBi

Potential Sources of Interference

It is important to understand that the two major frequency bands used by WLANs, the 2.4-GHz and the 5.2 to 5.8 GHz bands, are "shared" bands. In other words, WLANs do not have exclusive rights to these bands. Technically, any unlicensed frequency band is shared, but in this case the sharing is among fundamentally incompatible and interfering sources.

For example, the band of frequencies that we use in WLANs, 2.400–2.500 GHz, is part of the Industry, Scientific, and Medical (ISM) band, recognized by international agreement and operating in areas such as North America, Europe, and Asia. Unfortunately, this is a new "secondary" use for these frequencies. In the United States and around the world, frequency usage is allocated to specific groups of users for certain allowable purposes. Many frequency bands have multiple allocations to support the user communities. Such is the case with the 2.4-GHz ISM band. It shares a secondary allocation with several other so-called services or permitted uses. With one exception, these secondary allocation users are not required to be licensed and are expected to put up with interference from any other user.

Typical ISM uses, besides the "industrial" use of WLANs, include medical instrumentation, such as RF devices that heat tissue in the course of treatment; industrial devices, such as equipment used in manufacturing processes; and microwave ovens. In addition, there is a primary allocation 2.390–2.450 GHz for amateur radio users. In the case of this primary allocation, the allowable power levels are higher, and the users must be properly licensed Amateur Radio Operators. This band is used primarily for amateur TV, although other uses are permitted.[8]

In addition to these interference sources, you can expect to have potential interference from other low-power legitimate users, including Home RF networks, Bluetooth networks, and 2.4-GHz cordless telephones. All these are perfectly legal devices that operate in 2.4–2.5 GHz. All of them can interfere with your 802.11 network operation, although the timing of their use and the nature of their signals may make them more of an inconvenience than a serious problem.

Another WLAN allocation, comprised of the three 5-GHz bands (usually referred to as 5.2, 5.3, and 5.8 GHz) are actually part of the U-NII band plan of the Unlicensed National Information Infrastructure in the United States. The actual bands go from 5.150 to 5.250 GHz, from 5.250 to 5.350 GHz, and from 5.725 to 5.825 GHz. These bands also share allocations with other radio services, and will begin to experience similar interference as the 2.4-GHz band as soon as those uses increase. As with the 2.4-GHz band, the 5.8-GHz U-NII band segment has a similar overlapping allocation for amateur radio at 5.650–5.925 GHz.

We cover more information on dealing with those other networks and interference sources in subsequent chapters. For now, just keep in mind that these interfering devices are out there. For the most part, they will cause periodic or intermittent problems for your WLAN. These problems may be experienced as slow speeds, frequent re-tries, and brief outages.

---

[8]For a complete description of the American Radio Relay League band plans for these allocations, refer to the following link: http://www.arrl.org/FandES/field/regulations/bandplan.html#2300.

Wireless Internet in Every Room

Chapter Highlights

- Wireless Internet
- Installing W-NICs
- Installing Access Points
- Connecting to the Internet
- Using Antennas to Boost Range
- Project: A Simple Wireless Home Network

In this chapter, we will build our first wireless LAN for use in your home or perhaps for a simple wireless connection in a small office. This will be a wireless-only LAN, whose primary purpose is to connect your laptop or desktop PC to the Internet. In addition, it will connect two or more of these PCs in a wireless peer-to-peer network.

Because this is such a simple project, we will refer to it as a simple home WLAN, even though it can be used in an office, too. Keep in mind that this is just a basic building block for the more complex networks we will cover in later chapters. In fact, that is exactly what all wireless networking consists of: combining relatively simple structures to create more useful configurations. As long as we understand the basics of WLAN operation, which was covered in Chapter 2, we can build on those principles to create wireless networks of whatever size we need.

Wireless Internet

Our primary purpose for the simple home WLAN is to connect to the Internet from any location in the house. With this simple WLAN you will be able to connect to the Internet from any PC in any room without the use of network wires. If you have a laptop PC, you really will have portability. This is the point when you can begin surfing from your easy chair!

This project will assume that you already have a connection to the Internet through a cable modem or an ADSL line (often just called DSL). These connections provide "high-speed access" to the Internet. The term *high speed* is a relative one. This type of wired high-speed access can run at speeds of up to 1.5 Mbps. However, a speed of about 800 kbps is commonly quoted, and some connections may run as slowly as 128 kbps.

Most of these high-speed access connections have different upstream and downstream speeds. That is, the speed you can upload to the network is generally slower than the download speed. A really, really fast typist can barely hit 100 bits per second (remember typing speed is quoted in words per minute[1]), and most of us only infrequently send e-mail with large attachments that would fully use the upstream bandwidth. This type of access is called *asymmetrical*, which is the "A" in

[1]A very fast typing rate is 120 words per minute: 120 w/m × 5 characters/word ÷ 60 sec/min × 10 bits/character = 100 bps.

ADSL. By the way, DSL stands for digital subscriber line. You, the bill payer, are the subscriber to the phone company. Both DSL and cable modems normally have asymmetrical digital streams, with about a 4:1 or even an 8:1 differential. A symmetrical DSL line is available in some areas and is appropriately called SDSL (or sometimes HDSL for high-speed DSL).

There are lots of claims and counterclaims offered by the DSL and cable modem service providers. Some of the claims are that one high-speed service is faster, safer, or doesn't slow down like the other one. In reality, there are speed (actually throughput) variations throughout the day for all these services. Throughput may depend on time of day and be fairly predictable (such as when the kids get home from school and "hit the net" instead of the books). However, the speeds of both types of high-speed service will depend on the amount of other traffic that is being handled by that particular service at that particular time. The bottlenecks are simply at different locations. Whatever high-speed service you have, it will certainly be many times faster than a dial-up phone link.

Please do not confuse our WLAN project with wireless Internet access (Figure 3.1). Wireless access is a relatively low-speed connection you can make to the Internet through the cellular telephone network. If you have a cellular phone or a PDA with wireless Internet capability, you have this type of wireless access. You may be able to get a cellular wireless access modem for your laptop or connect through your cellular phone. However, you will get only speeds of 19.2 kbps or so. There is also a wireless (radio) service that uses microwave frequencies to provide Internet access to areas that do not have DSL or cable modem service available. The speed of this service is generally similar to ISDN at 64 to 128 kbps. It links to your home or office through a point-to-multipoint radio connection and requires antennas with line of sight to the service tower. This is a fixed connection to a single PC or a router. There is no moving around the house with a laptop. Our home WLAN project will give you much higher speed access to the Internet and the ability to roam through your house or office at will. You will essentially be your own wireless Internet access provider. There are no airtime charges and you get your "access" for free![2]

[2]With apologies to Dire Straits and MTV.

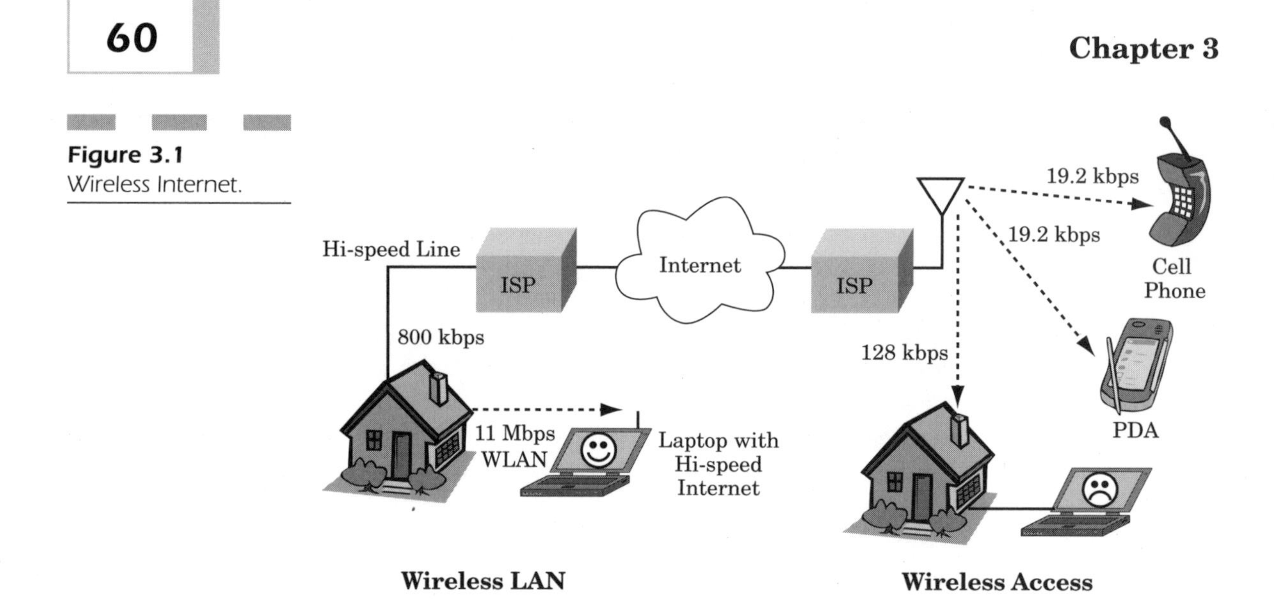

Figure 3.1
Wireless Internet.

Installing Your W-NIC

In this section, we are going to install a typical wireless network interface card (W-NIC). We cover the common elements of this installation, including the installation of the software drivers. The installation sequence is essentially the same, regardless of the brand of W-NIC you use. As a matter of fact, you will encounter more variation due to computer brand and OS than to the brand of W-NIC.

Before we get started, it is important to get a few basic ideas in mind. First, as with all option cards, the W-NICs are static sensitive, which means that static electricity can harm or kill a W-NIC, if you are not careful (see Figure 3.2). A few simple steps will take care of most problems. For example, you should make sure that you are not a walking lightning bolt of static electricity by first touching a metal desk, chair leg, or other object near your work area before you touch the PC card. Then you can simply touch the laptop case (a chassis-grounded metal part, such as an unpainted portion of the metal case is fine) and then touch the W-NIC. You should also store the W-NIC in a conductive plastic bag, such as those that all adapter cards are packaged in. In fact, your W-NIC came packaged in one. You didn't throw it away did you? Well, you can find another one. These bags are usually gray, dark brown, or sometimes pink, as opposed to clear, and often have a loose-weave, cross-hatched design printed in dark lines. It is a good idea to place the W-NIC on top of the static-suppression packaging while you are getting ready to install it into the PC.

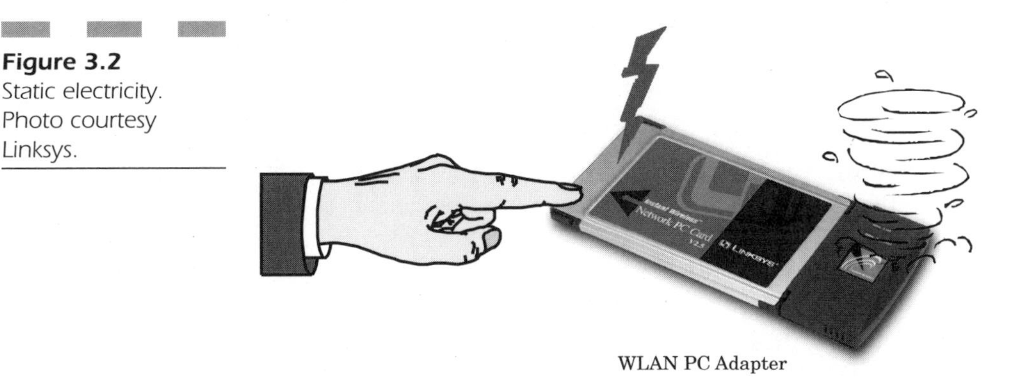

Figure 3.2
Static electricity. Photo courtesy Linksys.

If you use a W-NIC with a laptop computer, you may find it necessary to remove the card when you pack it up in the carrying case. It is tempting to tuck the card into one of the pockets of the carrying case, but be careful if you are not using a case that is specifically intended for laptop computer storage. Regular briefcases may use synthetic materials that can carry static charges. Often the laptop carrying cases are constructed with static-suppression materials in the interior compartments. If you think you might have a static problem in your case, it would be best to place the W-NIC in a static-suppression bag before you install it.

If you think this static problem is a joke, you certainly must live near the seashore or in an area with lots of humidity. Drier climates, such as the mountains, desert locations, or indoors in the north in the winter, produce the very low-humidity conditions that are ripe for large static electricity discharges. In some areas, static electricity is so bad that just reaching for the door handle can be a challenging experience. Realize that the voltages that must be present for that spark to leap from your hand to the doorknob are on the order of 5 to 10,000 Volts! Although the current is low, such voltages will permanently destroy the very tiny circuits inside your PC or W-NIC, so beware and be cautious.

The second basic idea is that most W-NIC manufacturers use the PC-card format (also called the PCMCIA[3]), as shown in Figure 3.3. This

[3]The term *PC card* is more ambiguous than PCMCIA (Personal Computer Memory Card International Association) because *PC* has many meanings. For those of us in the electronics industry, a PC card or board is any printed circuit board. PC also refers to personal computer. It would have been somewhat less confusing had the vernacular been PCM or PCC, but PC it is, so deal with it.

interface allows relatively small accessory cards to be added to the laptop computer without opening the case (as one must do with a desktop computer). The PCMCIA interface, in most cases, allows a card to be inserted or removed with the power on, without harm to the card or computer. (However, the OS software may get angry!) Follow the instructions from your laptop manufacturer, or the instructions for your OS. In general, you can insert a PC card without first turning off power to the laptop. If the card has never been installed, you will immediately be prompted for the appropriate installation disk (Windows 95, 98, ME, or XP). If it is recognized as a previously installed option card, the appropriate driver software will be loaded and the card will be "started."

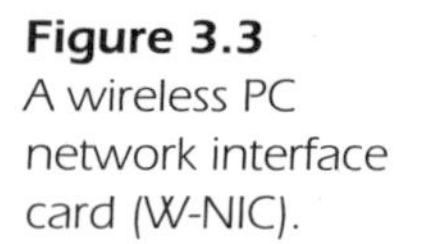

Figure 3.3 A wireless PC network interface card (W-NIC).

You should always shut down the card before removing it while the laptop is on. You can do this by using the applet provided by the manufacturer. This applet is normally located in the Control Panel folder (Start/Settings/Control Panel). It is a good idea to place a shortcut icon for this applet on your desktop, if you frequently use this function.[4] If you forget to "shut down" the card, and remove it without turning the laptop power off, you most likely will get a message reminding you to "down" the card first. However, in some circumstances, you may interrupt a process in software that will "hang" the machine. Your only

[4]To add a shortcut to the desktop for the PCMCIA manager applet, open Control Panel (Start/Settings/Control Panel) and then use the right mouse button to drag the icon for the applet onto the desktop. When you are prompted, choose "Create Shortcut."

option may be to turn the laptop off and restart, for which you will receive another computerized scolding. The best practice is to always shut down the card before removing it with the laptop running. Of course, you can always remove or add a PC card when the laptop is off.

Installing a W-NIC in Your Laptop

A step-by-step process for the installation of your W-NIC is given in the project, Simple Wireless Home Network, later in this chapter. At this point, we will discuss some of the issues involved and the general steps in the installation.

Once you have purchased the wireless adapter, the next step is to install the PC card in your laptop. You may want to review the tips on choosing a W-NIC and AP that we covered in Chapter 2. Most WLAN manufacturers have a "starter kit" that includes both an AP and one or two W-NIC cards (Figure 3.4). Remember, the best way to be assured that the W-NICs and the AP are totally compatible is to buy both from the same manufacturer. After you have your basic system up and running, you can certainly experiment with other brands of W-NICs. There is a fairly wide price variation among WLAN components that are intended for SOHO use. You may want to start with a well-known brand to get the quality, support, and warranty. Once you are comfortable with your installation, you may add WiFi™ cards from a different source, with little or no apprehension that the card and AP might be incompatible.

Figure 3.4
Example of a WLAN starter kit.

Choosing WLAN components that are compatible with your operating system is also important. For example, you may have a computer that uses an NT-based OS, such as XP. These OSs are often incompatible with drivers intended for Win 9x OSs. Be certain that the W-NICs and the AP have driver support for your OS.

To install your W-NIC in a typical laptop computer, you should unpack the card, locate the driver disk (which may be a 3.5-inch diskette or a CD-ROM), and place the disk nearby your laptop PC (Figure 3.5). Remember what was said earlier about static electricity. Keep the card in its static-free wrapper until ready for use. If you do not plan to keep the card in your laptop all the time, be sure to observe static-handling procedures each time you insert or remove the card. Because laptops rarely have metallic cases, you can use the metal connector shells that are along the back of the laptop. Be sure to touch the metal shell first, not the tiny metal pins in the connector. They are directly connected to circuitry in the computer and would subject the internal components to damage, if you sparked to them.

You also might want to be sure you are not bearing any nasty static charges before you touch the laptop or the W-NIC. A good way to do this is to sit down at your workspace (walking around is the way we usually pick up static electricity) and briefly touch a metal chair or table leg to discharge yourself before you touch the PC. Static charges dissipate gradually through the air, so these objects will be at or very near ground potential. If you walk around a lot while you are working on or near your computer, you could wear static-discharge shoes. (Or just walk around in stocking feet not synthetic socks, or even barefoot. Not that we want to get exactly third-world here, but this would eliminate most of the static problems.)

In industrial situations, the workers wear a grounded wrist-strap to discharge any static buildup. The strap is a high-impedance connection to ground, so there is no chance to get harmed by the strap if the workers should inadvertently touch a live wire or a component that has a substantial voltage above ground potential. Gee, maybe if computers get any more high tech, they will have us running around in dust- and static-free "bunny-suits" such as those used in semiconductor (chip) processing plants.

If your computer has a modular option for CD-ROM or floppy drives (as many laptops do), be sure that the appropriate drive is installed to match the driver disk that came with the W-NIC. Go over the instructions that came with the W-NIC. Most accessory cards for laptop PCs have very similar installation routines, which we will cover next. However, if there are any differences for your W-NIC, you certainly should follow those instructions rather than the general guidelines here.

Figure 3.5
Installing a W-NIC PC card.

You usually can insert a new PC card after your laptop is turned on, but a more conservative approach is to install it with the power off. This is particularly good the first time you install the card because it is a little clumsy to get the card seated properly while avoiding the eject button. It would not be good to have the card inadvertently pop out with the laptop on. Practice inserting the card and ejecting it a few times with the laptop off, so that you can get used to the process.

Installing a W-NIC in a Desktop PC

Typically, the W-NIC will be installed in a laptop computer when total wireless portability is needed. However, you can certainly install a WLAN connection to a desktop computer to avoid having to run network cabling. The installation sequence is much the same in a desktop computer, with the addition of an adapter card for the W-NIC (typically a PCI-adapter card; Figure 3.6).

To install a W-NIC in a desktop, you should follow the manufacturer's specific directions. This will normally call for installing the adapter card first, and then the W-NIC. Most PCs recognize the addition of a PCI card and will automatically configure the internal system settings, such as the interrupt (IRQ) number.

The first step is to do a normal shut down of the computer, turn the power switch to OFF, and remove the power cord. (Many newer computers do not really turn off, but merely "sleep," with one electronic eye open to start up at the slightest touch to keyboard or mouse. This can be a problem if you happen to be inside the computer when it starts up!) Next, using the same static-electricity precautions we covered in the prior section, remove the portions of the computer's case needed to install an option board. In general, this will be the left panel when viewed from the front (that's the starboard side for you captains and mates). There may be one or two small Phillips-head screws on the portion of the side panel that wraps around the back of the case. Occasionally, you will need to unlatch the front panel on the computer and unscrew or release the side panels there. If you have any questions, refer to the upgrade instructions for adding option cards that came with your computer.

Figure 3.6 Installing a PCI adapter in a desktop PC.

Don't be a contortionist. Professionals always put the computer over on its side, so that they can clearly see the motherboard and the option card slots. Get plenty of light (a flashlight or strong room light) and grab a magnifying glass or your best glasses if you cannot see everything clearly. Mis-seating or improperly placing an option card can cause real problems. In addition, you might knock something loose, so be careful to touch only the components you need to, and to make certain that any cables you do accidentally bump or intentionally move out of the way are properly re-seated before you finish and close the case. It may be best to carefully unplug everything from the back of the chassis and move it to a table or countertop for dis- and reassembly. Also, it is a very good idea to cover your work surface with a cloth or some cardboard to avoid scratching it (the surface, not the PC).

Now, prepare the case for the adapter card. The adapter card will probably be a PCI-interface style, as the older ISA adapters are becoming rare and many recently made computers only have PCI expansion slots. You will need to remove the oblong metal panel that covers the appropriate slot in the rear case. Sometimes, this is just a folded metal insert that can be pushed outward until it pops free. In an older chassis, it is part of the rear case stamping that is literally hanging by a metal thread, not stamped totally through. In this case, a slotted screwdriver can be used to pry the remaining stamping back and forth until it breaks free. Be careful of the edges of these chassis slots (and the metal case, as well), because many manufacturers don't seem to understand the value of deburring the sharp edges, and you can cut yourself badly.

Locating the correct rear panel slot for the adapter card is a bit of a bother. Older ISA-bus cards (these are relatively long, dark, dual-edge connectors) have the connector on one side of the appropriate slot, and PCI-bus cards (shorter, white, dual-edge connectors) have the connector on the other side. If you are not careful, you will remove the wrong rear panel insert, and the things are really devilish to get back in (impossible if you have the break-off kind). Hold the option card near the slot you wish to use, and you will clearly see which side of the slot the connector should be. You will need to pick a slot that is not too difficult to access after the case is reassembled, so try to avoid a slot that would be covered by cables and such. You may want to move one or more of the other cards to get the best placement for your W-NIC adapter. Remember, the radio signals for the AP must be able to easily reach the W-NIC antenna, if the internal antenna is used. The signals may have difficulty getting through if they have to penetrate a mass of cables at the rear of the computer, so a location toward the top of the case (when set back upright) is best.

After you have decided where to place the adapter card and prepared the chassis slot, remove the card from the static-guard package and install it into the option slot. None of the manufacturer's manuals will tell you this, but the easiest way to do this is to get the card roughly in position over the correct connector and then place the lower tab of the metal header (the part that will cover up the slot hole in the chassis) slightly down into position between the motherboard and the chassis. Now align the front edge (opposite end from the header) of the card with the PCI connector and press it down so it begins to go into the connector. Now firmly rotate (like a lever) the card the rest of the way down into the connector so that the folded-over part of the metal header comes down flush with the rear chassis shelf. Do not press overly hard because you do not want to damage the motherboard. Finally, rock back again with slight pressure on the front edge (probably where the heel of your hand is) to ensure that edge is still properly seated in the connector.

To finish installation of the adapter card, use the appropriate screw to secure the option card header to the rear chassis. In some cases, it may be necessary to remove the card, bend the header very slightly, and replace the card to insert the screw without partly pulling the opposite end of the card out of its connector. Screws that come with PCs have two general thread "pitches," which we technical types call "fine thread" and "coarse thread." Use of these terms avoids metrification and metric terminology as much as possible, and it really works. Most adapter cards use the coarse thread screws. You can easily see which is which when you look at the assortment of screws and connection components that came with your computer. It is a good idea to keep these items handy, perhaps taped securely inside the chassis in the plastic bag. You never can tell when you might need a good threaded component.

Now it is time to reconnect the computer and load the drivers. You can probably reassemble the case at this time, if the design of your adapter card allows the PCMCIA W-NIC card to be inserted after the case is closed. Almost all cards do. Just reverse the process you used when you disassembled and disconnected the computer. (Try not to have too many parts left over.) Remember that you will still have to insert the W-NIC card, so you might leave the computer chassis so that you can easily get to the back to plug in this card, if it cannot be done with the case closed.

After you have all the cables and accessories properly reconnected, you can power ON the computer. If you have a Windows 95, 98, or XP OS, the computer will recognize the new card on boot up and will ask for the appropriate software driver for the new adapter card (Figure 3.7). In some cases, the appropriate driver may be available to the OS and you will see a message: "Windows has located new hardware and is installing the appropriate drivers." If you have Windows NT or 2000, you will have to use the Add Hardware wizard in the Control Panel. If your manufacturer recommends it, you may need to install drivers after Windows has completely booted. If so, use the disk provided by the manufacturer and then reboot your computer (generally not needed for NT or 2000). Always make sure that the OS is totally happy with your hardware and software before adding more components, and a reboot is the easiest way to be certain of this. If you have any problems, your adapter card manual should have detailed troubleshooting instructions. You can also go to the System applet in Control Panel, look at the Device Manager tab, and note any devices that are not properly installed (designated by a question mark in a yellow circle).

Figure 3.7 Windows' Add New Hardware Wizard.

After a problem-free power-off reboot with the new adapter card, it is time to add the W-NIC to the PCI adapter. First, do a system "shut-down" and completely power the computer off. Next, install the W-NIC into the PCI adapter. Most of these adapters are generic PCMCIA adapters, so they have the same mount and eject button as the laptop's PCMCIA port. Now reconnect the power cord, turn the power supply switch to ON, and, if necessary, press the power button on the front panel.

The desktop PC will recognize the addition of the W-NIC card (Windows 95, 98, ME, XP) and prompt you for the driver software. Follow the instructions above, as for a laptop PC. If you are running Windows NT or 2000, you will have to use the Add Hardware wizard available in Control Panel to install the W-NIC card and its drivers. To complete the installation, you will often need to reboot the computer. If your W-NIC or AP requires the installation of additional software, you may install it at this time.

Installing a USB W-NIC

The Universal Serial Bus (USB) interface offers a really exciting way to add WLAN capability to a computer. In the case of a desktop computer, it is particularly useful, because you can avoid all the complexities of the desktop installation described in the previous section. Although all WLAN manufacturers do not offer the USB option, all of the major suppliers do, so you should have no problem finding a USB W-NIC (Figure 3.8), if that is your choice.

Figure 3.8
A USB WLAN adapter.
Courtesy Linksys.

The installation of the USB W-NIC proceeds much the same way as the laptop installation. The main difference is that you will be installing a standard USB rather than a PCMCIA device. In fact, your USB W-NIC adapter may really be a USB-PCMCIA adapter with a regular W-NIC PC card included.

When you install a USB W-NIC device, you are really installing two components, the USB adapter and the W-NIC. That means you will have to install driver software for both components. Some manufacturers offer very specific instructions on the order of installation of these components. With plug-and-play OSs, such as Windows 95, 98, and XP, you may be able to plug in the USB adapter and let the OS prompt you for the installation disk (also called a setup or driver disk).

Many manufacturers seem to take particular pleasure in hiding not only the drivers on the disk but also the instructions for installing the drivers. Experience has shown that even top-tier manufacturers often delight in placing obtuse or even totally wrong instructions in their consumer packages. This seems to be particularly true of USB installations, possibly because they are multistep installations and procedures vary greatly depending on the specific OS version and hardware configuration. Frankly, we think that these reasons are no excuse at all, certainly for after-market options that the consumer is expected to install.

That being said, we will give you some tips on installation that may help you through the maze if you run into one of these poorly documented USB units. We have placed these tips in a Tech Tip, USB Ports and Driver Installations, so you can skip over them if you have an easy installation.

Tech Tip: USB Ports and Driver Installations

This Tech Tip describes the basic requirements for a USB installation and some fairly straightforward techniques for getting around the installation problems with troublesome driver disks. Although these techniques include information addressing USB issues specifically, we also cover some of the generic driver installation problems that apply to all driver software.

First, do not try to install USB hardware on a Windows 95 system. With the last release of Windows 95 and some releases tailored for original equipment manufacturers (OEMs), such as all the computer names you know so well, it became possible to add the USB interface adapter drivers to the OS. However, it is not well supported in Win 95 (which is tech talk for "it doesn't work very well, and the manufacturers will not help you"). In addition, many of the drivers recognize the presence of Win 95 and simply refuse to install. It can create lots of problems, so the best practice is simply to upgrade to a newer OS. Although it is not the latest, an upgrade to Win 98 from Win 95 is fairly painless and trouble free and you will like your computer much better—plus Win 98 fully supports USB.

Second, you will need a USB host adapter. The newer computers have from two to four of the USB ports built in. If your computer lacks USB (but has the right OS) you can buy a two- or four-port

USB adapter (a PCI or ISA option card) and add it in. Use the same installation instructions as for the desktop W-NIC installation in this chapter.

Third, you need a powered USB adapter port. The USB specification for USB ports powers connected USB devices directly. This is really handy, because you can get accessories that do not need external power adapters (sometimes called "wall warts" because of their unappealing appearance). You also don't have to find a place to plug more accessories into electrical outlets. However, there are limits. Most powered USB ports are limited to about 500 mA, which is a lot for most external devices. However, if you use a USB hub to expand to more USB ports, you may not have enough extra juice to power your USB W-NIC, because some USB hubs are line powered (that is draw their power from the USB port on the PC). These unpowered hubs are fine to expand USB connections to printers and other devices that have their own sources of power, but will not work with so-called high-power USB accessories such as a USB W-NIC. If your USB W-NIC requires line power, the solution is to connect it directly to one of the PC's USB ports and use your unpowered hub on the other port.

Fourth, if you have problems finding the proper USB driver on the installation disk, check this out. Most drivers are delivered on CD-ROMs that include a variety of drivers for virtually every OS in current use. This may make it difficult for the manufacturer to use the autoloading feature of some OSs. Although this is really not that difficult to do, these same manufacturers invariably have more difficulty writing clear instructions. Many times, you will have to browse the files on the CD to find the proper driver. One convenient way to do this is to choose the Specify Location option for the driver installation wizard. When the Found New Hardware wizard pops up, choose Have Disk or Specify Location. Enter the drive letter for your CD and click Next or OK. The wizard will look at the top directory of the disk, probably not find a driver, and prompt you (again) to put in the correct disk. Click the Browse button, choose the drive for the CD (usually D:), and scan through the directories on the CD. Usually, they will be clearly named with the OS, such as Win98, NT, Win2000, XP, MAC, etc. Highlight the directory that matches your OS and click OK. The wizard will find the driver loader, which is normally an ".inf" file. Highlight the driver that matches your W-NIC model. Click OK again, and the driver will install.

Upgrading Your W-NIC Firmware

Every W-NIC runs internal software (or firmware[5]) that controls the operation of the W-NIC. Periodically the manufacturer will issue updates to the code. The updates will fix minor problems and, in some cases, may offer additional features. This is particularly important for changing the behavior of the card as new standards emerge. This feature also allows you to put off "planned obsolescence" a little longer.

To get the current firmware updates, go to the W-NIC manufacturer's Web site and download the latest version, along with the latest versions of the drivers and the management software. Installing the firmware update is simple. You should have your W-NIC installed and operational in the computer first. Then you install the firmware as you would any other program. In general, you will double click the firmware update file, which will launch, find the existing W-NIC card, detect the card's current firmware, ask you to confirm the update, and then perform the update. The firmware programs are usually very small (tiny in comparison to most of your software), so they install very quickly.

You can then use the W-NIC manager program to check the status and firmware revision of the card. At this point, you can also do a self-test or, better yet, a signal-strength test to the AP. This will confirm proper operation of the card with the updated firmware. The new firmware is held in a special type of nonvolatile memory called Flash-ROM, so that it does not have to be reloaded until you want to do another update.

Setting Up Your WLAN AP

The next step in installing your WLAN is to install your wireless AP (Figure 3.9). In some simple installations, it is convenient to set up your AP temporarily so you can check out your W-NIC installations because the W-NIC in your laptop or desktop computer will need something else to network with. You can move the AP to its permanent location later on.

---

[5]*Firmware* describes replaceable software associated with specific hardware. Virtually everything from an appliance to a rocket ship has "programmable hardware" components. In a PC, the internal boot-load chip contains firmware called the BIOS, or basic input/output system. Many option cards also contain miniature operating programs (or machine code) that are retained with power off. W-NICs typically contain their firmware in a special type of rewriteable memory called Flash-ROM, so we can download updates to the code but do not have to reload the code every time we turn the computer on.

Figure 3.9
A typical wireless access point (AP).

The AP is the equivalent of a wireless hub. Most WLAN starter kits will contain one AP and one or more W-NICs, or you can buy all the components separately. Remember that the AP is just another wireless node on the WLAN as far as the wireless side of the network is concerned so all APs will include the functionality of a W-NIC, just like any connected device. Some APs include an internal W-NIC, whereas others have a PCMCIA slot into which you plug an ordinary W-NIC. In fact, just such a card may lie hidden inside the case of the AP. The reason for this is simple economics and the burden of Federal Communications Commission (FCC) type acceptance (or the equivalent in Europe or Asia). It is much cheaper for a manufacturer to use the same W-NIC in the stand-alone (to be placed in a PC) and the AP configurations than to develop another unique circuit just for the AP.

In the days when APs were very expensive, the modular W-NIC configuration also allowed the user to replace the W-NIC with new technology cards when they emerged. This was much less costly than replacing the entire AP. However, new APs are far less expensive, although they still may be two to three times the cost of a W-NIC.

Temporarily Installing Your AP

The initial installation of an AP is pretty much "plug and play."[6] That is, the AP, even if it is designed for a plug-in W-NIC, comes from the manufacturer set to operate. Of course, initial operation is in a default configuration, so you may need to change the configuration for your particular network. Also, the AP will likely come with all security features turned off, so the user will have no difficulty installing. Just keep in mind that you will need to put in your own passwords and enable the security features after you have checked out the operation of your laptop W-NIC connection using the AP's default settings.

You can pick a temporary location for the AP that is near the computer in which you installed your W-NIC. It is also convenient to have the DSL, cable modem, or router nearby, so you can check out the Internet access. Eventually you will need to connect the AP to the access equipment anyway, so why not make it convenient. If you pick a temporary AP location that is near your W-NIC–equipped computer, the wireless signal will be strong, and you can set aside any concerns about positioning the AP and associated antennas. However, you should be aware that maintaining at least a few feet of separation between the AP and the W-NIC minimizes any chance of signal overload between the units. A distance of 5 to 10 feet is good (approximately 2 to 3 m).

Your AP will likely come with an installation disk. This is not really to install the AP, but to install AP management software on your laptop. This management software will have the ability to connect to the AP and allow you to configure such things as the network name (unique to 802.11 operation) and the level of security. It will also allow you to pick the operating channel, do testing, and in some cases assign an IP address, so that the AP can be managed remotely.

In addition to configuration from a wireless node (such as your laptop with the management software), some of the more sophisticated APs can be configured over the Ethernet interface or even over a serial port. In a few cases, these APs may need to be preconfigured to operate properly, or they may need to be reset to factory defaults to be initialized. Some of these considerations are discussed in more detail in Chapter 8, as they are probably not going to face the average small or home office user.

---

[6]*Plug and play* without the hyphens (Plug-and-Play) refers to the ability to "just plug it in" and it will work rather than the Windows PnP™ feature that allows options to be automatically configured. Think of it as plugging in a new TV or radio. You don't have to do much other than unpack it and plug it in.

For a step-by-step guide to installation, see the project Simple Wireless Home Network, later in this chapter. In short, to install the simple AP, you need to plug it into a power outlet, connect an Ethernet cable to your Internet-access digital modem, hub, or router, and install the management software on a computer that has a W-NIC already installed. With a little bit of luck, and favor from the god of defaults, you will be connected to the AP *and* the Internet. If you are not sure that you have a *wireless connection*,[7] you can use the W-NIC manager application that you installed with your W-NIC to scan for an AP. Another tip is to make sure you can connect to your access modem on a conventional wired Ethernet cable before you try to connect through the wireless connection. Most access devices will use automatic IP address assignment to assign the IP address to your PC. The AP functions just like a bridge, so it passes all traffic between the wired LAN[8] and the wireless nodes, and the IP addresses for source and destination are transparent.

One word of caution: you need to have networking (specifically, TCP/IP) installed on your PC to make any connection to the Internet, whether on a direct cable or through your WLAN. That is why it is a good reason to check out a wired connection first. If the wired connection will not work, it is certain that the wireless connection will not. If you have never installed networking on your computer, see the Tech Tip, Installing TCP/IP Networking.

Tech Tip: Installing TCP/IP Networking

If you have never installed TCP/IP networking on your PC, you will need to do so before you can operate your wireless network or any wired network. Here is a simple set of actions you can take to see whether TCP/IP is installed and configured correctly to operate in a typical home network. These instructions will work with wired or wireless networking, if you have a Windows 95, 98, ME, NT, 2000, or

[7]*Wireless connection* is sort of an oxymoron, but we will go with it anyway. The data connect, even if the wireless devices don't.

[8]In our simple home network example, this LAN consists of just two devices, the AP and the access modem (an ADSL modem or cable modem). In some cases, the access modem may have a built-in router and a small (four-port) hub, or there may be a simple router/hub between the cable modem and AP. Technically, each side of the simple router would be a tiny, two-node LAN.

XP system. XP has some wizards that aim to simplify the process, so the language is a little different, but the concepts are the same.

1. Check the status of your networking

To install IP networking on your computer, you will need to install the network adapter driver and IP. In general, when you install a network adapter, the adapter is added to Windows networking, but you may need to manually add the IP protocol. To check these items, open the Start/Settings/Control Panel window and double click to open the Network icon (or, if the Network Neighborhood icon is on your desktop, right click it and choose Properties). If you find that you have the adapter or protocol already installed, you can skip that step (although you might want to see that the TCP/IP properties are set up correctly).

[NOTE: You must have your Windows Setup CD to make any changes to the Network setup. If you do not have it available, DO NOT proceed.]

On the Configuration tab, be sure that your network interface card is in the list (scroll down, if you need to). If it is not there, you should choose the Add button, select Adapter, click Add again, and follow the instructions. Try not to press Enter during this process because it closes the Network window and prompts for the Windows CD. We have more to check and configure before we do this. Your NIC will need to be physically installed in the computer and you will need to have the software driver disk available.

Next, you need to scroll down the list of installed components on the Configuration tab to the protocols. You should see TCP/IP -> ReallyFast Ethernet Adapter (actually, the name of your NIC). If not, you will need to add the protocol with Add/Protocol/Microsoft/TCP/IP. *Remember not to press Enter or click OK at the very bottom of the dialog.* After you have added the protocol, select it in the Configuration view, and double-click the protocol or choose Properties. You can select each tab in turn to ensure that the choices are set to "automatically assign an IP" (DHCP) and that the Gateway and DNS are provided by DHCP. In rare instances, you may need to enter these values, if they are not provided automatically through your ISP. In general, you can disable WINS, unless you need access through an NT domain. This is beyond the scope of our topic here.

Now, you can finally hit OK. You will be prompted for the Windows CD and you will be asked to restart Windows.

Windows 95, 98, ME, NT, and 2000 have very similar networking setup screens, but Windows XP is different. On XP, you should begin with Start/Control Panel/Network & Internet Connections (or you can open My Computer). Next make the following string of choices: My Network Places; View Network Connections; click on the connection; View Status of This Connection. The Details (click arrow to open) panel will show the IP address and possibly "Assigned by DHCP," which is OK if it is in one of the expected ranges in item 2, below. To check/change the TCP/IP properties, highlight Properties and Internet Protocol (TCP/IP) Properties. All these windows must make it very drafty inside an XP box. All of the selections and the process of adding adapters and protocols are the same, but the appearance of the selection and information panels has changed.

2. Check for a proper IP address

If you have a Win 9x system, the easiest way to do this is to connect your computer to the access modem (DSL or cable modem), reboot (skip the reboot if you were already connected the last time you rebooted), and run WINIPCFG (Windows IP Configuration). To run this applet, go to the Start Menu (the button at the lower left of the task bar), choose Run, enter "winipcfg," and click OK. You can get the same information on NT/2000/XP systems, but this handy utility is missing. One way is to go to a DOS window and type "ipconfig."

In WinIPConfig's pull-down box (which will, annoyingly, probably display the PPP adapter, by default), choose your Ethernet card (something like "RealFast 340a PCI NIC"), and look at the IP address. It should show one of the following:

1. An Internet address assigned by your Internet service provider (ISP), which may be a DHCP assigned dynamic address
2. An automatically assigned nonroutable DHCP dynamic address (which comes from your own router)
3. A bogus "filler" address

Case 1 means that you are directly connected to a cable or DSL modem with no router function and the ISP has given you a network address or that you have a "static address" also assigned by the ISP. In either event, you are fine, as long as you can connect to the Internet through your browser. For Case 2, your address will start with 192.168.x.x or 10.x.x.x, which are the two more common ranges of "nonroutable" addresses that can be used on any local network inside a router. Typically, your router will assign an address ending in ".100," ".101," ".103," etc., so the entire address would look like "192.168.1.101." Sometimes, the "10.x.x.x" (said "ten-dot") address range is used. If you see an address that begins with one of these ranges, you know everything is fine.

Case 3, the default Windows "filler" address, is bad news. It begins "169.254.x.x" (the x represents randomly assigned addresses), which indicates that Windows IP services did not automatically get an address assigned by the router (and no static address was preassigned), so it assigned one anyway. Although it looks like a valid address, the address actually belongs to Microsoft and is bogus because it will be on neither the ISP's subnet nr on your subnet (unless you are a Redmond dude). It means that you did not connect, possibly because of a problem with your network cable, router, or cable/DSL modem. Sometimes, you can click the Release button, and after a short delay, the Renew button to get a proper IP address assigned. If you successfully release the address, the IP address will normally display 0.0.0.0, and when you renew, you will see a valid address other than the 169 one.

You must have a valid address to proceed. If you do not, you will have to troubleshoot your connection, check your network properties, and possibly reinstall some of the software and drivers. Use the network troubleshooter in the Help application for Windows. The strongest recommendation we can make before implementing a wireless network is to get IP Networking and your Internet connection working first in the wired world and *then* install your W-NIC hardware and software.

You can complete the temporary installation of your AP by connecting it to your router (or directly to your cable/DSL modem, if you are not using a router). This involves connecting an Ethernet cable between the two devices. With a router, you will often have a modular hub, which is sometimes built into the router. If you use a stand-alone hub, keep in mind that the router should be plugged into the hub at the cross-over connector. Some hubs have a small switch, say for Port 1, to select between the normal and cross-over interfaces. Other hubs actually have two physical connectors for Port 1 (with one marked "X" or "1X" to indicate the cross-over). This cross-over port is sometimes called the "Concatenate" port. See Figure 3.10 for a diagram.

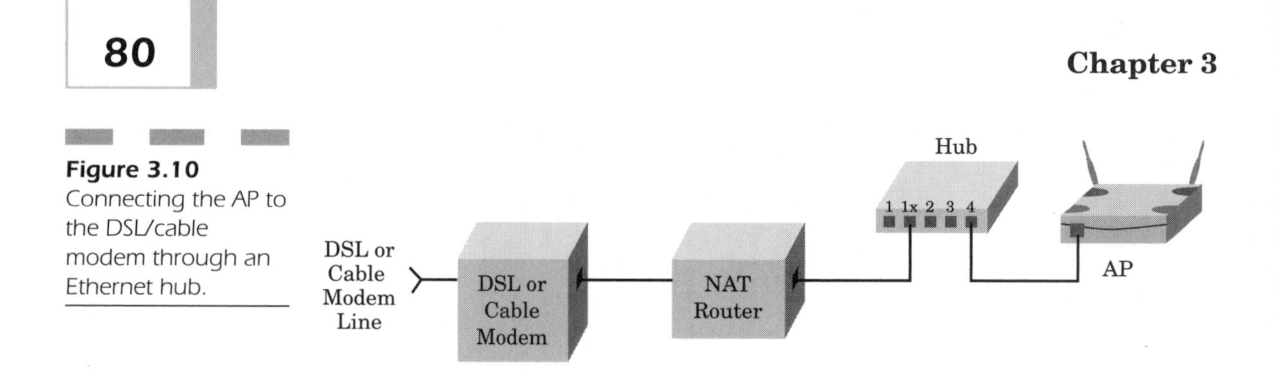

Figure 3.10 Connecting the AP to the DSL/cable modem through an Ethernet hub.

Placing Your Wireless AP

Now that you have installed your W-NIC, temporarily installed your AP, and demonstrated that both can work together through your router, you will need to permanently locate your AP. You should consider the locations of your cable or DSL modem, your router and hub, and the proposed locations where you will want to use your wireless connection. Remember, the AP has to connect via cable to the access device, and it will need power. You will probably want the AP in the same area as the router/hub/modem, unless you are into running LAN cables around your home or office.[9]

It is ironic that installing a wireless network connection does involve some cables. At a minimum, you must cable the network access router to the AP. In most SOHO installations, you will have other PCs, servers, or printers as part of the wired network to which you attach wireless networking. All network cables should be good-quality modular Ethernet cables, usually called Category 5 (or perhaps 5e) patch cords. If you have to go quite a distance from a hub to the AP, keep the standard rules in mind: the total cable length must be shorter than 100 m (329 feet) from hub to device, including all patch cords/cables. This is really a lot, unless you are running cable up, down, and around a building.

Pick a location that has easy access to power. Most APs are not weatherproof and will have to be mounted inside, so you may want to test how well the AP's wireless signal will reach through the walls (or window) to the outside locations, such as patio, pool, or workshop. If you have the type of AP that will accept an external antenna, you can mount the AP inside, near the wall where the antenna cable will come in. Remember, cables longer than a few feet have really terrible losses at 2.4 or 5.2 GHz, so you want to get the AP as close to the antenna as possible.

[9]The reader is referred to *LAN Wiring,* 2nd ed.

Also consider the fact that certain objects may interfere with your wireless signal. Think of the light bulb analogy in Chapter 2, in which metal objects, such as refrigerators, HVAC units, air ducts, and file cabinets act as large metal shields to your wireless signals. The farther you get the AP and the W-NIC from these objects, the better chance your signal will have to get through. You also may have problems with metal screens, metallic (reflective) window films, foil-covered wallboard, and metal doors. In addition, cement or block walls may partly block (or attenuate) the signals.

If you have any doubts about the proper location for your AP, you can use the link tests in the next section to help you determine the best location for it. You may need to add a second AP to cover problem areas. You should set the second AP to a different channel, so you don't interfere with the first AP. The W-NIC in your PC will scan to acquire the best signal from whichever AP is strongest. However, part of the process involves authentication and connection to the W-NIC by the acquiring AP. At this time, the standards for roaming between APs are in development, so you probably won't be able to walk (or otherwise move) between the two APs, without reinitializing the connection. This is not the best of possible worlds, but at least you can get fairly good fixed coverage over a wide area by using multiple APs. If the cell-to-cell AP roaming is an issue for you, try to find a brand of AP that implements a multiple-AP protocol. Although this may be a proprietary solution, at least it will get you through until the standards are fully implemented. Chapter 8 discusses these details and AP-to-AP bridging. Eventually, we will all be able to walk around from AP to AP, even in public areas, with constant connectivity to the Internet.

Testing Your AP/W-NIC Installation

The management software that came with your W-NIC includes the ability to do diagnostic and performance testing on both the W-NIC and the link to the AP (Figure 3.11). This feature will tell you whether your W-NIC is installed and working properly, at least as far as the OS software is concerned. In addition, you will be able to check for a connection to an AP and probably will get an indication of how strong the signal is. However, you will probably not get any idea of how the AP is performing.

To check out the AP, you can open the AP management program. In the Windows environment, if there is no desktop icon, you can choose Start/Programs/name_of_your_AP/AP_manager_program (the latter two

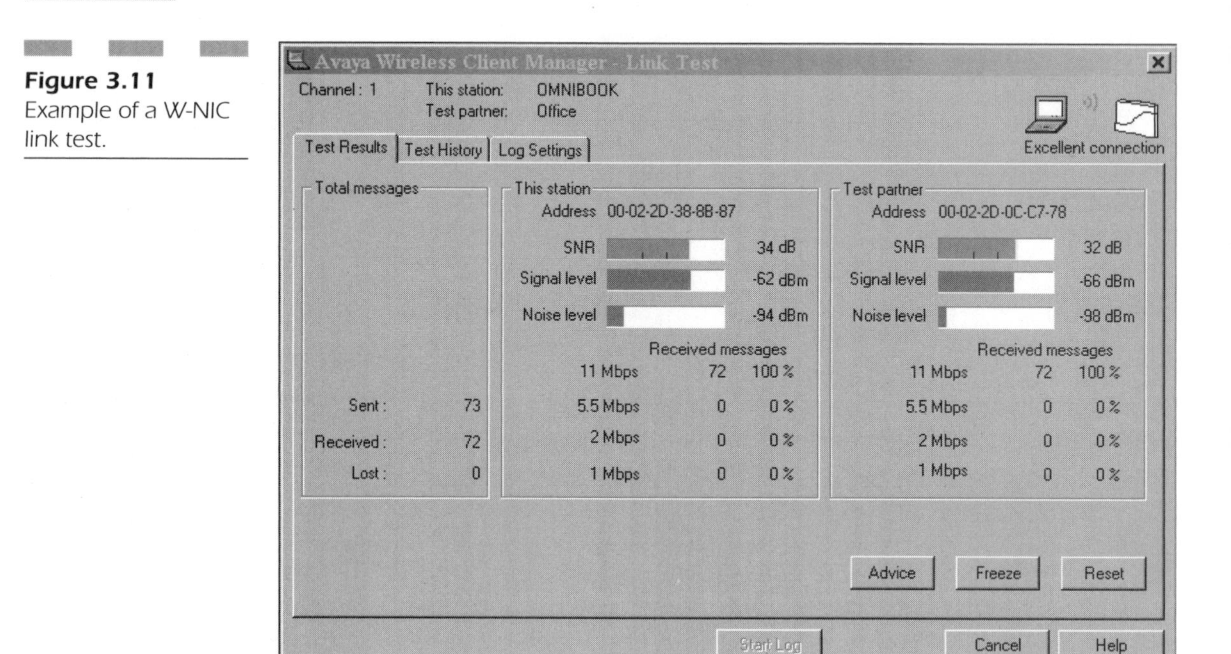

Figure 3.11 Example of a W-NIC link test.

items are specific to your brand of wireless hardware). Once the AP manager program opens, you can check for a connection to the AP and then choose the "link test" to check out the entire wireless connection (Figure 3.12). This is nice, because you get both sides of the connection. Remember, there are two devices involved, the AP and the W-NIC, and you need to know the received signal strength for both devices to get a complete picture of the connection. A good AP management program will let you see both sides: how the AP is receiving and how the W-NIC is receiving. The best management programs actually display a graph and a numeric value showing the receive strengths at both ends.

Occasionally, you will have an AP that needs to be manually set via a direct connection. The APs that require this type of setup are generally those that are intended for large office/campus use, perhaps with multiple W-NICs and external antenna capabilities. If you have one of these APs, the factory default configuration may not allow you to control the AP through a wireless connection with your AP management software. If this is the case, you will need to make the proper direct connection through a direct-connect 10/100 Ethernet cable or perhaps a serial connection.

Security for your AP is one more item we need to consider. In addition to the wireless management connection for your AP, there is often a possible connection through IP, the Internet Protocol Suite. This connection

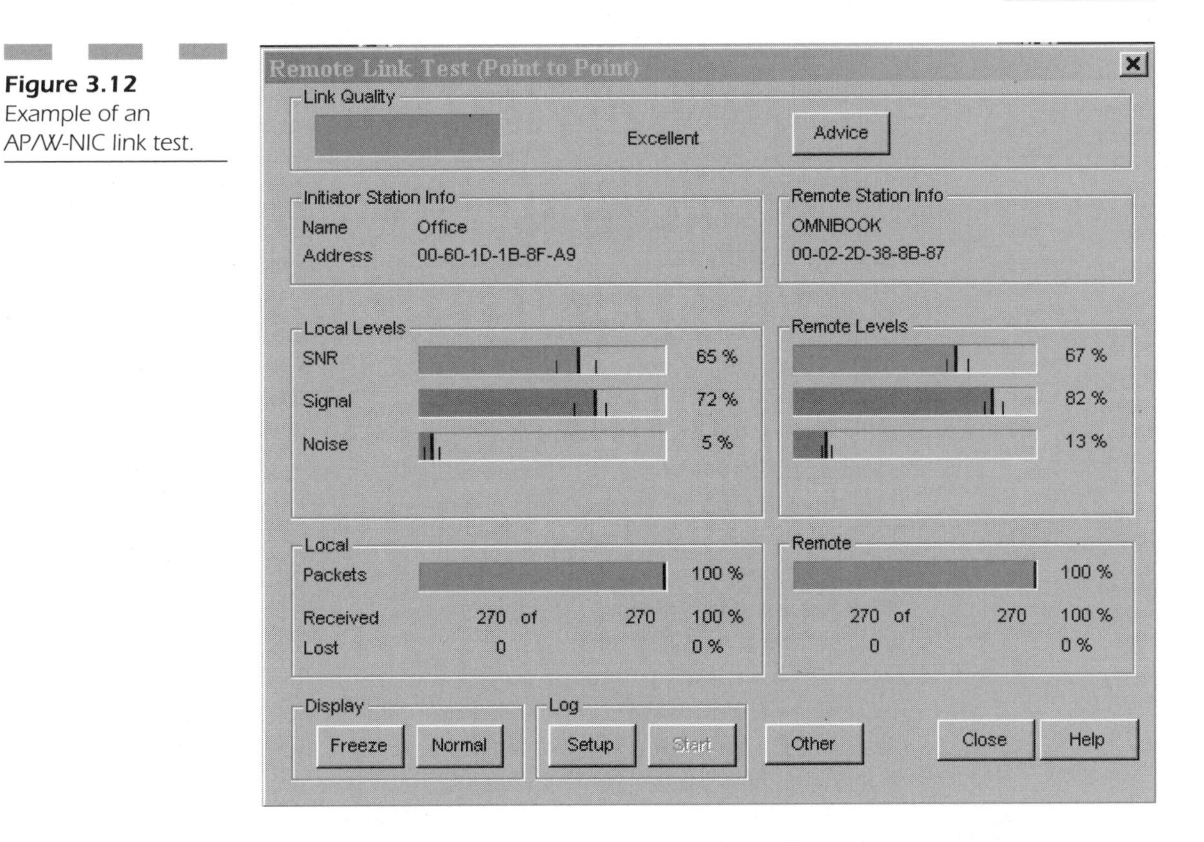

Figure 3.12 Example of an AP/W-NIC link test.

is called Telnet and is a convenient command-line interface that is common to many routers, servers, and workstations. Telnet provides a convenient ability to remotely manage APs and is a significant benefit in larger wireless networks. However, it also poses a potential security threat because anyone with the actual IP address of the AP can get into the "control panel" with Telnet and thereby gain access to your private network. So if you have an AP with Telnet capability, regardless of whether you use it, you should disable Telnet or password-protect Telnet access.

In addition to Telnet, some APs may use Simple Network Management Protocol (SNMP) or even Hypertext Transfer Protocol (HTTP), the protocol that runs the Web. You are using HTTP if you use your web browser to manage your AP. Most of the combo wireless Internet access routers (an AP and a DSL/cable modem router in the same box) use HTTP for management. As soon as you get one of these units set up, you should change the password from the default, if there is any password set at all. This will be the password, and perhaps user name, that is required for setup or administration of the AP/router. Although the default password may seem unique to you, be assured it is not. Every hacker out there (as well as the good guys) knows the default passwords that come with every popular brand of hardware. So, be safe: change it right now!

Now, we can have a little fun with the link test applet. Once you get the graphical/numeric display on your W-NIC or AP link test application, you can move the AP and/or the laptop around to see how the signal changes. See what happens when you turn the laptop around, say 180° on a table top. You should see the signal change as the case of the laptop gets in the way. Now pick up the laptop and turn it up and down, side to side, and observe the graph indicator. (This is much more difficult with a desktop PC, and we do not recommend it, even for professionals.) If you have a W-NIC or AP with small rod or rabbit-ear antennas, try moving them around for best signal. Notice that some antenna positions that maximize signal strength in a particular orientation of the laptop will cause a poorer than normal signal in other orientations. Pick the antenna position that works best in several orientations, unless you can leave the laptop stationary.

Now, if we were someone else, we might recommend one other brief experiment. You know that the human body is not totally transparent to the WLAN frequencies; actually it's pretty opaque. Of course, you must stay at least 1 inch from the antennas of these devices to totally comply with the FCC RF exposure guidelines (similar to guidelines in most countries) so we would never suggest that you do this at home. However, if you *were* to set up the link test and then grab the antenna with your hand (or put the palm of your hand over one of the imbedded antennas), you would likely see a dramatic signal decrease. Of course, we do not in any way suggest that you should do this, but the results are neat. Also, this can give you an intuitive understanding of some of the issues in RF signal propagation. For that matter, you could also see what the signal difference is when you hold the laptop computer and turn your back to the AP, so that your body partly shields you from the signal.

Making Your Internet Connection

You can use your wireless network to gain access to the Internet in every room of your home or office. For that matter, because wireless networks are not limited to fixed wiring topologies, you can get at least some limited Internet access outside your home or office. Closed-in areas, such as porches, patios, swimming pools, gardens, or hot tubs, will probably have sufficient wireless signal strength to reach your AP. Of course, you should take the proper precautions to protect your computing equipment in these locations, so that humidity or heat from direct sun does not damage your laptop.

The project later in this chapter will give the step-by-step process for installing your simple wireless home network. Here, we discuss the features and some of the issues for your wireless Internet connection.

To make your wireless connection to the Internet, you will need one or more laptop/PCs with W-NIC cards installed, a wireless AP, an Internet access device (such as a cable modem or a DSL modem), and, of course, an Internet access line. All these items are shown in Figure 3.13. You may use a router, a router with a built-in Ethernet hub, or a combo AP/DSL/CM/router for your access device. In addition, many SOHO networks also have a small, wired network to which the AP and DSL/CM access device are connected. We cover such a network in Chapter 4, *Home Office and Small Office LANs*.

Figure 3.13 Components of a wireless Internet connection.

The Internet access line should be a DSL or cable modem line for a high-speed connection. Most of these so-called high-speed lines will give you at least a 768/128 kbps downstream/upstream connection. You will commonly hear of data rates of around 800 kbps for these lines. This represents a typical download speed from the Internet (actually from your ISP) to your computer (in this case, your DSL/CM access device). Most high-speed connections limit the upstream (or upload) speed for technical and practical reasons. Among these is the fact that our typing and clicks on Web browsers are at a very, very low data rate, and upload of e-mail is infrequent and brief.

Of course, if you are running some Web servers in your little (?) home network, this is bad news, because it will make your servers seem slower to the outside world than they really are. But it is good news to the ISP because they really don't want you doing that from your home network, over a connection whose overall bandwidth they intend to share with a lot of other home users. If you want high-bandwidth *to* the Internet with a symmetrical Internet connection, you will have to get a T1 line or perhaps an HDSL connection. These other types of lines cost significantly more than simple DSL or cable modem lines, but they are worth it if you need the extra bandwidth. You will also need special, more expensive, access routers with these other lines.

A lot has been said about faster and faster WLAN speeds. WLANs are often compared with wired LANs, most of which run at 10 to 100 Mbps. Consequently, WLAN technology did not really take off until speed was increased from 1 and 2 Mbps to 11 Mbps in IEEE-802.11b (and to 54 Mbps in 802.11g). That's a 10:1 speed increase, which is very significant if we need to support heavy local client-server applications over a local wireless network, for example, from a server in our small office network to our laptop. You can see that if the "high-speed" Internet connection maxes out at only 800 kbps (which is only 0.8 Mbps), then having anything over that for our wireless connection into the Internet is not of much value. This is illustrated in Figure 3.14. Still, there is "spin" value to the advertising, so we run to the higher and higher WLAN speeds. Realistically, some DSL and cable modem connections can get up to about 1.5 Mbps, but the fact of the matter is that the infrastructure and backbone Internet connections of the ISP, coupled with the size, speed, and load of the host servers have the greatest effect on your page-download speed.

As more streaming media content is used on the Internet connections of the world, having at least 800 kbps over each and every link will become more and more critical. We will begin to see the next generation of access, perhaps using fiber to the curb (FTTC) Ethernet connections to our homes and offices. Once this happens, the value of having a WLAN speed of 11 or 54 Mbps will be realized. Because we know this is coming and may, for some areas, come over the next three or four years, you will be better prepared if you go for the higher WLAN speeds now, so you will be ready for the inevitable progression of technology.

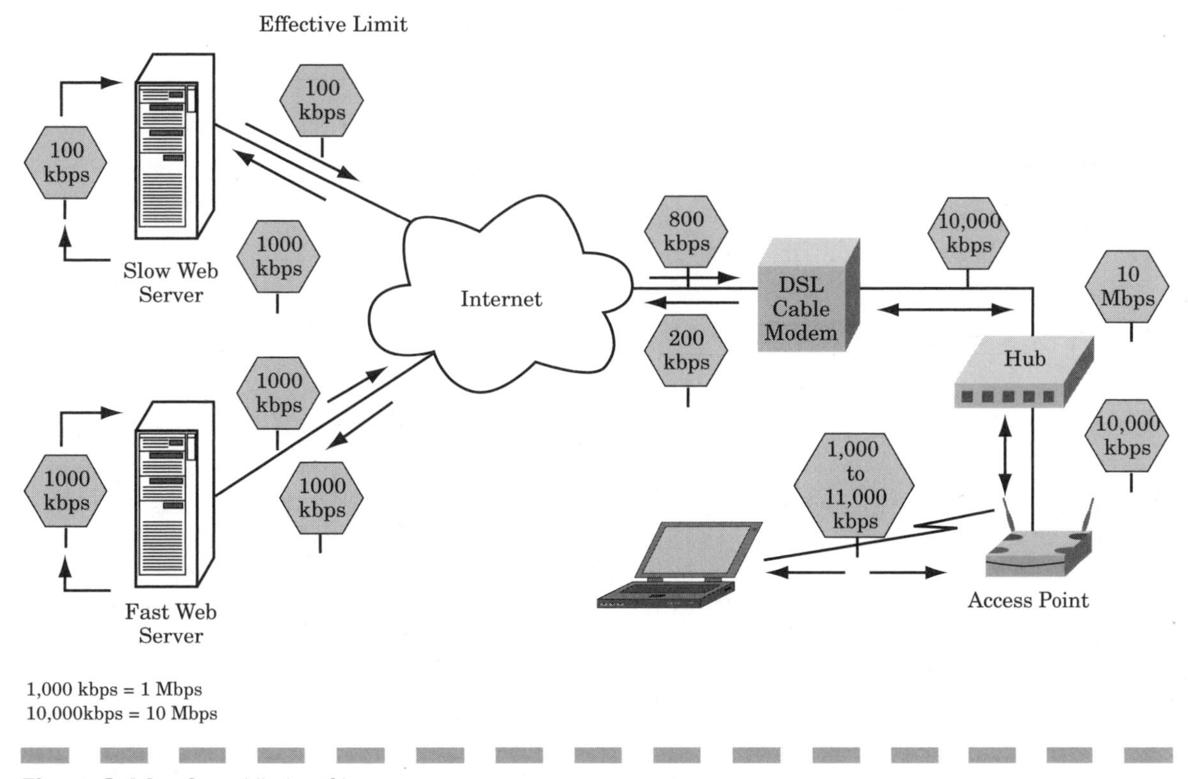

Figure 3.14 Speed limits of Internet access.

To make your wireless Internet connection work, you can first test the wired connection directly from your PC to the access device. If you do not have a wired Ethernet card on your PC, you are probably safe to connect wirelessly first. This is particularly true if you have a combo wireless router.

Using Antennas to Boost Your Range

WLAN manufacturers will quote ranges of 50 to 500 m. This is about 150 to 1500 feet. Typically, a larger range is quoted for "outdoor" as opposed to "indoor" use; this attempts to convey that the absolute maximum usable range will be outside of any buildings, for example, in an open field. Unless you plan to use your wireless network that way (outstanding in your field), you will probably have doors, walls, and furniture in the space between your AP and your W-NIC–equipped computer. In an average open office, you may get as much as 150 feet (50 m) of range, whereas at home (with lots of walls, floors, or air conditioning ductwork in the way) you might get merely 50 feet (not meters!) of range. Fortunately, with many wireless components, you can add external antennas to boost your range.

Antennas for the AP

The easiest place to add an antenna, and the most practical one, is at the AP. Many APs are equipped with a connectorized factory antenna or two. This is illustrated in Figure 3.15. This AP model allows the connection of two antennas, which may be shared for transmit and receive, or may be split between those functions. Many APs will have only a single antenna connector. In either event, you can add a length of cable between the AP connector and the antenna to place the antenna in a more opportune position.

Figure 3.15 AP with external antenna connectors (shown with one antenna removed).

In addition, special-purpose directional antennas are available to increase the strength of the wireless signal in certain directions, such as to your favorite lounge chair near the pool. These antennas are available from third parties or the W-NIC or AP manufacturer. The antennas are designed specifically for the WLAN frequency of operation and the fairly wide bandwidth of the channel. If you add antennas to your installation, you should be sure that they are for 2.4-GHz IEEE 802.11 operation (or, for the proper frequency band, 5.2 to 5.8 GHz if you are using IEEE 802.11a equipment). Antenna cables will have considerable losses at these frequencies, so you should keep them as short as possible. Consider relocating the AP, rather than using a long cable, if that choice is feasible. The Ethernet Category 5e (or Category 6) cable that connects the AP to the network is capable of running 100 m (329 feet) with virtually no loss of data. The multi-gigahertz signal to the antenna will lose strength every foot of the way. By a poor choice of location and cable length, you can easily sacrifice all of the advantage of a higher or better antenna location.

An external antenna may be a handy way to get WLAN access outside a building or through an interior wall, without having to provide a weatherproof enclosure for the AP. If you do mount an antenna outside, you should consider whether you need lightning protection. This is particularly important if you mount the antenna at or near the roof of a building. A few million stray electrons can wreak havoc with the sensitive circuitry in the AP.

Theoretically, it makes little difference at which end of the path you place the external antenna, at the AP or at the W-NIC. On the one hand, it will be much easier to place an antenna on the AP, or to move the AP to a better location. On the other, it may be the chassis of the PC, or perhaps a metal desk or cabinet, that is blocking the signal between the two. This is particularly relevant to W-NICs in desktop PCs because of the placement of the W-NIC antenna (at the very rear of the metal chassis) and because the desktop cases are more often placed on the floor (behind a desk or cubicle wall).

Antennas for Your Laptop or PC

You may be able to add an external antenna to the W-NIC in your laptop or desktop PC. However, only a fraction of the W-NIC manufacturers allow an external antenna connection. Those that do, such as the one shown in Figure 3.16, often have the connector hidden by a tiny plastic plug. Removing the plug reveals the RF connector. The external antenna must be attached with a cable that has a matching RF plug.

Several types of external antennas are available to enhance the range to your PC. Some of these antennas have a weighted base, with a verti-

Figure 3.16 W-NIC with a connector for an external antenna.

cal antenna element, so you can place them high on a cabinet or shelf to improve performance even more. The tiny coax cables that typically connect the W-NIC and the antenna are adequate to run 4 to 6 feet, but for longer distances, you should use larger low-loss coaxial cable.

You may need an adapter to translate between the tiny W-NIC connector and the much larger low-loss cable connectors. These larger cables typically use what is called a "Type-N" connector. You may also find a connector called a "TNC" and the types of connectors often used for cellular radio antennas. Some miniature connectors are variations of the SMA connector. However, some of the W-NIC manufacturers use a specialty or custom connector that is not widely available. The purpose may seem to be to encourage the user to purchase antenna options from them, but the real explanation is that all these WLAN components contain RF transmitters that must be certified by the governing authority. In the United States, this certification is done through the FCC. By using a special connector, the manufacturer can ensure that only antennas that are properly certified to meet FCC requirements are used.

Project: Simple Wireless Home Network

This first project shows you how to install a basic WLAN and connect to the Internet. This project is appropriate for an initial wireless project and may be combined with the wired network components in the second project in Chapter 4. If your goal is to gain Internet access for one or more laptop or desktop computers, and you don't need to share printers or files, then this project will fit your needs perfectly. This first project is also suitable for experimenting with WLAN technology, testing wireless range, and evaluating throughput.

Although we call this a "home network" project, it is certainly suitable for a small office. In many cases, you will use WLAN connectivity for business, and it works just as well in home and office locations.

The primary benefit of our Simple Wireless Home Network is to allow you to connect to the Internet at your home or small office, without being tethered to the wall with wires. If you have a laptop computer, you have noticed that it is not truly portable unless it can be unlinked from the network connection. Otherwise, you have a cumbersome network cable that must be stretched across the room or from room to room. In addition, you may even damage your laptop PC (or yourself) if you accidentally catch your foot on the strewn cable. You really cannot get portable until you are wirelessly connected.

This project also allows you to connect to many public-access WLANs in hotels, airports, and coffee shops. Of course, you will have to subscribe to a compatible service to gain access in most of those locations. But once you are connected, you will have instant access to your e-mail, financial sites, or entertainment sites on the Web.

In addition, we cover installing the PCI adapter card in a desktop computer. This adapter allows you to have wireless access to your full-size PC, eliminating the need to run cables throughout the house or office.

What You Will Need:

1 DSL or cable modem line
1 DSL or cable modem
1 Wireless AP
1 W-NIC, 802.11b wireless PC adapter card

Optional:

1 PCI adapter for W-NIC
1 W-NIC, 802.11b wireless PC adapter card (for PCI adapter)

Project: Simple Wireless Home Network

1 SOHO router for DSL/cable modem use
2 Category 5e 10/100BaseT patch cables

Project Diagram

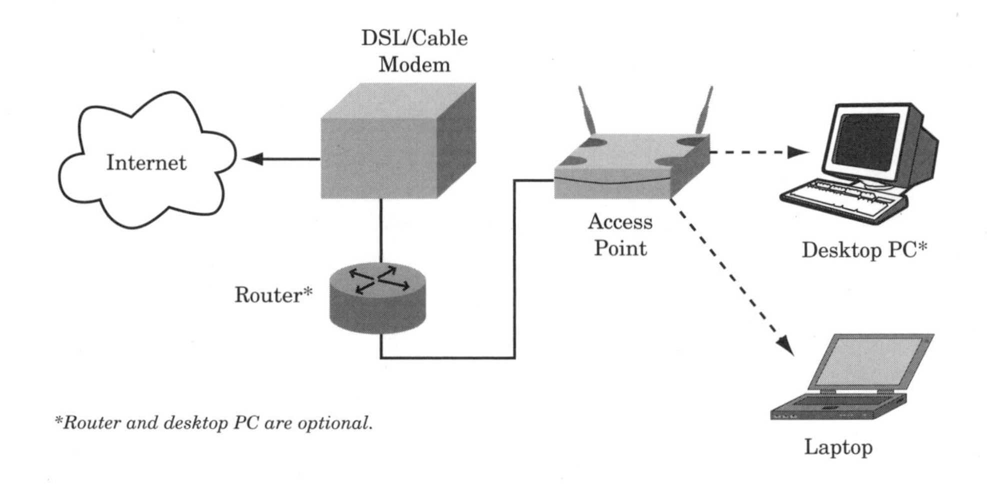

Project Description
For this project, we will first construct a simple WLAN using one wireless AP bridge and a laptop PC with one W-NIC. Then we will connect the AP to a DSL or a cable modem line for Internet access. As an option, we will then install a PCI adapter card in a standard, desktop PC, and add a W-NIC to make that PC wireless, too. In most cases, two or more wireless PCs will need to connect to the DSL/cable modem through a router, so we will add the router as an option.

Step-by-Step
Step 1: Install a W-NIC Card in the Laptop PC

For this step, you should prepare the laptop computer for installation of the W-NIC card. We will assume you have a Windows-based OS. It is best to shut down the computer from the Windows desktop rather than pressing the power (or sleep) button. Many laptops save their current operating environment, including the state of the desktop and all open applications, to disk, so that they can return quickly to the fully awake state. In this “suspended” mode, the computer is not really “off,” just asleep, with all the applications parked temporarily on the hard disk. You need to make sure that you have a complete shutdown, with no problems, before installing the W-NIC.

To shut down Windows from the desktop, go to the task bar (at the bottom of the desktop), choose Start/Shut Down, and in the Shut Down Windows dialog choose Shut Down and click OK. A full, successful shutdown occurs when the screen displays “Now it is safe to turn off your computer” in large red print and promptly shuts off. In some cases, the laptop will shut down immediately after completing its shutdown sequence. If you encounter a problem here, the laptop will most likely “hang” with a Windows Logo screen displayed. If the condition does not clear after a couple of minutes, you will have to press the power button again, or in some cases, hold the button down for several seconds. You should then immediately restart the laptop, allow it to complete its startup sequence, and attempt a shutdown again. In some cases, this may need to be repeated two or three times until Windows clears whatever condition was causing it to hang.

NOTE: You *must* have your Windows Setup disk (e.g., Windows 98 CD) available on CD-ROM or loaded on the laptop to proceed with steps 1 and 2.

Next, unpack the W-NIC, locate the installation disk, and be sure that the CD or diskette drive is installed, as appropriate for the type of installation disk. Some laptop PCs have an option slot for the CD or diskette drive, and do not have both available at all times. Remember to keep the W-NIC in the protective static-discharge wrapper until you are ready to install the card. Go over the specific instructions that came with the card. We include general steps here, so that you will have the information in case the manufacturer’s instructions are not clear.

After you have shut down the PC and readied the W-NIC materials, you can begin the installation of the card. First, locate the Type II (or Type III) PC (PCMCIA) slot(s) on your laptop PC. In most cases, the W-NIC will require a single slot, but it may interfere with other cards that have been installed, if your PC allows more than one PC card at a time. If the W-NIC interferes with the other card or you only have one slot, remove the other card. It is common for some analog modem or Ethernet cards to use two slot positions to accommodate the connectors. You will not need these other cards while you are using your WLAN, and you can replace them when you need to use the modem or a wired LAN.

The W-NIC will have a connector at one end and the plastic antenna housing (possibly with fold-out antennas) at the other end. With the card oriented upward (with the manufacturer's logo and card name facing upward), insert the connector end of the card into the PC slot. It should slide in easily and the connector should seat with a little firm pressure. PC slots normally have a small square button or lever at the right-hand side of the inserted card. This button ejects the card. It is a push–push mechanism, which means that you push it once to cause it to spring out and push it a second time to eject the card. If you accidentally push the button in, you will have to push it a second time to get it back in place, which unfortunately ejects the card. It is a bad idea to do this with the PC on, so practice pressing the button a couple of times to get the hang of it.

The fairly long W-NIC, with its protruding antenna housing, can usually be removed by just grabbing the card and pulling, but best practice is to always use the eject button. This avoids damaging the card or misaligning the antenna housing. You should always try to place the W-NIC in the top slot if you are using a second card in a dual PC slot. That will give you better wireless coverage than being partly shielded by another card and its cables.

Now, turn the laptop on. Windows will detect the addition of the new W-NIC and will bring up the Add New Hardware wizard. (These instructions vary slightly for other Windows 9x versions. NT and 2000 systems will have to add the adapter manually.) Choose Next, then Search for the Best Driver, and Specify a Location. Enter the drive letter for your CD-ROM drive or diskette drive, as appropriate. You may need to browse to a specific directory on the driver disk to locate the proper drivers for your operating system. For example, you may need to select D:\Win98 for the Windows 98 drivers. See the instructions that came with your W-NIC for specific instructions, if you have trouble. The wizard will locate an appropriate driver, or a list of possible drivers, if there is more than one. Select the proper driver and press Next.

Windows will proceed to load the W-NIC drivers and other needed drivers from the Windows Startup disk. After the setup is complete, you will be notified and should click the Finish button. You will be offered an opportunity to restart your computer to complete the setup.

Project: Simple Wireless Home Network

Step 2: Install TCI/IP Networking

To use the card with the Internet, you will need to install Internet protocol (IP). This is part of a protocol suite, which is a fancy term for a collection of networking programs for data interchange. The Internet protocol is a part of TCP/IP networking. (The TCP stands for transmission control protocol.)

NOTE: You can skip most of this step, if your computer is setup for TCP/IP networking. However, you will need to change the Properties pages for your W-NIC adapter, if they differ from those recommended here.

First, open the Network applet (Start/Settings/Control Panel/Network). From the list, select your W-NIC adapter (it was added to this list when we completed Step 1). Click the Add button, select Protocol from the list, and click Add. Chose Microsoft in the Manufacturers list and then highlight TCP/IP in the Network Protocols list. Choose OK (on this dialog, not on the underlying Network applet).

Second, select the newly added TCP/IP protocol (with the → pointing to your W-NIC card's name) from the list on the main Network dialog. (Scroll down if you need to.) Click the Properties button. Now verify the following settings on the respective Properties tabs:

1. IP Address: Obtain Automatically
2. WINS Config.: Disable (unless you log into a domain)
3. Gateway: (Blank)
4. DNS Config.: Disable DNS
5. Bindings: (The appropriate components are checked)
6. Advanced: (You may check Set as Default if appropriate)
7. NetBIOS: (Check Enable NetBIOS over TCP/IP if appropriate for your environment)

If you are not sure which setting to use, try the defaults first. The only one that will probably need to be as shown above is the IP Address, as the computer will gain its IP address from the DSP/cable modem company, unless you use the optional router.

After you have verified the Properties page, click OK on the Network applet, and you will be prompted for the Windows 9x CD. Insert the CD and click OK to continue (or the system may find the needed files on your hard disk). You will need to reboot the computer to finish loading the settings. The setup for NT, 2000, and XP is a little more complicated, and you should consult your system manual for specific instructions. In general, you will need to manually complete some of these tasks to add an adapter and enable it.

Step 3: Install the Management Utility for the W-NIC

At this point, you have installed the W-NIC and driver software and enabled TCP/IP networking. It is very likely that the default settings on your W-NIC are fine to initially connect to the AP. However, you will need to modify the default passwords and network name for security reasons and you will need to monitor the signal strength to the W-NIC from the AP. To do these functions, you should load the W-NIC management software that came with your wireless adapter card.

Installation instructions differ somewhat among the various WLAN manufacturers. We can summarize the procedures here, but you will need to look through your installation manual to get the specific instructions for configuring your particular brand of W-NIC.

Most utility packages start up automatically when you insert the CD in the CD-ROM drive. If it does not or your PC does not have Auto Launch enabled, you may need to start the setup manually. To do this, you can go to Start/Run and enter D:\setup (where D is the drive letter for your CD-ROM drive). You can also click the Browse button to look on the CD for your setup program.

Many generic utilities will try to install in the root (top level) of the C: drive. This is bad practice, because your directory list can fill up with all sorts of folders and get very confusing. It is better to force the utility to install itself in the Program Files folder rather than directly under C:. A very few utilities will require the actual eight-character (8 × 3) file name rather than the long name of Program Files. This name is "progra~1".

After you install the management or configuration utility, you should check the default configuration of your W-NIC. The following list shows typical default choices that are appropriate for most APs. In this project, you will be building a wireless network to connect through an AP rather than peer-to-peer between wireless computers. In IEEE 802.11 lingo, this is called *infrastructure* mode. The peer-to-peer setup is called an *ad hoc network*.

Project: Simple Wireless Home Network

Mode:	Infrastructure
Service Set ID (SSID):	(Default*)
Channel:	(Normally not an Infrastructure choice, as the card roams to find the AP)
Encryption:	(Default*)
TxRate:	(Default, probably automatic)

**Keep the default until you have linked with your new AP. Then you will want to change the Network Name (Service Set Identifier) and the encryption.*

By leaving these at the default settings, we allow you to initially install the network and verify basic operation before you change any settings that might cause difficulty. If you obtained different brands of AP and W-NIC, you should expect that the Network Names will be different, at the very least. The best idea would be to write down the default settings on the network card and, if needed, change them to match the AP's defaults. After you get the W-NIC installed and communicating with the AP, you can change the settings to whatever you want. Actually, you should change the SSID and the encryption from the default settings to make it more difficult for snoopers to monitor your wireless network transmissions.

Step 4: Install the Wireless AP Bridge

The next order of business is to install the wireless AP bridge. If you purchased the AP and the W-NIC from the same manufacturer, there is a high probability that both components are set with default configurations that will allow you to begin instantly to communicate. This is particularly true if you purchased the components as part of a "starter kit" for WLANs. However, if the two components were purchased at different times, from different sources, or you have reason to believe that the software and hardware revisions may be separated by time (and thus in version), then you may have to do a little configuration on one or both components to jump-start your wireless network. We cover both cases.

Our purpose in this step is to complete the wireless mini-network consisting of the AP, the laptop PC, and its W-NIC. Initially, we are not concerned about TCP/IP, just about the 802.11 link between the two devices. After the link is set up and operating, we can proceed to the next level and get the IP working. Then we can use IP to connect all the way to the Internet.

The first thing to do is to unpack the AP and plug it into the power source. Switch it on, if the AP has a power switch. You do not need to worry about connecting it to the DSL/cable modem (or optional router) yet, but if you already have those components functioning and feel brave, go ahead. Place the AP in the same room as your laptop PC (the one you installed the W-NIC in) and preferably near your DSL or cable modem.

Next, boot up your laptop with the W-NIC installed. After your laptop has completely booted and loaded the W-NIC drivers, start up your W-NIC configuration utility (installed in the last step). If everything is copacetic, the utility will indicate a "link" and show the actual received signal strength and link quality. Keep in mind that this is the W-NIC's utility, so it shows only the signal *from* the AP to your W-NIC, and not the other way.

At this point, you can load the AP manager utility on your laptop. The AP software will be on the disk that came with the AP bridge. The process will be similar to the way you installed the W-NIC manager utility.

Most APs may be managed one of three ways:

1. Over the wireless connection
2. From the Ethernet connection
3. Through a direct cable, serial, or USB

Whatever the method, you will need to use the AP Management/Configuration utility to configure, monitor, and option the AP. If your W-NIC manager utility indicates that you *do* have a wireless link to the AP and if your AP allows management over the wireless link, this is the easiest way to connect. Simply start the AP manager utility on your wireless laptop and, once a connection is shown, make whatever configuration changes you want to the AP. In some cases, you will need to manually set an IP address in your laptop because there won't be a connection to the router's DHCP server to assign an address.

Most APs have a default IP address already assigned, probably within the 192.168.1.x range. To avoid conflict with SOHO routers, many of which use the address 192.168.1.254, some AP manufacturers set their default address to .250 (that is, 192.168.1.250). You can choose any address within this range, other than the AP or router address. The address 192.168.1.101 usually works, if there are no other devices in the subnet. Just go to Start/Settings/Control Panel/Network and choose the Properties for your W-NIC's TCP/IP protocol entry (much as we did in step 2). Once your PC and AP are set to compatible addresses within the same subnet, you can configure the AP over the wireless connection. Some utilities also allow this simple method of connection over wired Ethernet, so you don't have to resort to Telnet or SNMP managers. Be sure to change the default configuration password for security.

After you are finished, be sure to reset the TCP/IP properties to "Automatically obtain an IP address," so you can pick up an available address from your router. This will avoid any possible conflicts with other PCs on your network.

Managing the AP over the Ethernet connection is much the same as over the wireless link. You have to be on the same subnet, and you will probably have to manually enter your laptop's IP address, if you are not given a compatible IP automatically from your router.

The third method is to use a serial or USB cable (whichever your AP needs) to connect. With the USB cable method, you definitely will have to install special software on your laptop, if that is the computer from which you do the configuration. However, unlike the other two methods, you do not need to set an IP address.

Step 5: Connect to the Internet

At this point, you should have a very fundamental WLAN that connects from your wireless laptop computer to the AP. Now, you need to connect through to the Internet. Connect your DSL or cable modem to the access line, if the installer did not already do this. The modem will have an Ethernet port to connect to a single PC or a SOHO router. If you have a router, connect the AP to the router using a Category 5e Ethernet cable. You should get link lights on both devices. (Some equipment refers to the link as the LAN.) If not, you may need a cross-over cable that reverses the transmit and receive pins, so that each device gets the proper signal.

Next connect the router to the DSL or cable modem using a similar Category 5e Ethernet cable. Again, you should get good link lights on both devices. As before, in some circumstances, you may need a crossover cable, if you do not get a link light. The modem may have come with a choice of two cables. (Careful: a DSL modem will include a special RJ-11–style cable that will plug into a 10/100 Ethernet port but will not connect the proper pins. This cable is for the DSL line connection, not the Ethernet one.)

The router will fetch a global IP address from the ISP and in turn the router will supply private IP addresses to your PCs. Once your wireless laptop is properly connected to the AP and the AP is connected to the router, your laptop will automatically fetch a private IP address from the router by connecting through the AP. It is as if the AP is a bridge between the wired router and the wireless laptop. No joke!

You may need to release and renew the IP address on the laptop to ensure you have a valid one in the private subnet's range. Remember, this happens automatically when you reboot the PC or manually through Start/Run/WinIPcfg.exe. If you do reboot, remember that the router needs a global IP first, so boot it first and the laptop second.

If you do not have the optional router, simply plug the AP directly into the DSL/cable modem with a Category 5e Ethernet cable. As before, you should look for link lights on both sides and replace the cable with a cross-over cable if necessary.

Now you should be able to access the Internet from your wireless laptop PC. If you have done everything properly and if your high-speed line is ready to function, you will be able to open a Web browser on your laptop and connect to your favorite sites for e-mail or entertainment over your new wireless network.

Troubleshooting

If you have problems, be sure your browser is not set to use a proxy server and go back through these steps. A good troubleshooting technique is to try to connect to each device in the string through its IP address. For example, from the PC browser, enter the AP's IP address as if it were a URL. (The configuration utility will do the same thing.) If you can connect to the AP, then try the router's private IP address. This will be the *Gateway* address shown in WinIPcfg. If you *cannot* reach the router, then there is a problem with your IP address, with the DHCP settings, with the AP, or with the cable to the router. If you *can* connect to the router, then try to go to the router's own gateway, which you can find from the router configuration screen. This is probably what you accessed when you browsed to your Gateway address. If you cannot reach this gateway, there is a problem with your DSL/cable line connection, and you should check with your provider.

However, if you get this far and can indeed reach the ISP's gateway, you probably have a DNS problem. DNS, or Domain Name Service, is the means by which a URL (Uniform Record Locator), commonly known as a Web site address, gets translated to the physical IP address of the Web site. For example, if you wanted to go to www.microsoft.com/whatever, the DNS that your computer points to would direct you to one of the main microsoft.com DNS servers, such as 207.46.138.20. That DNS would further refine your request until you reached the proper server for /whatever. If your DNS settings are wrong or missing, your browser can't look up the "microsoft.com" DNS IP address, so you won't get any content from that site or any other on the Web. You may need to manually set the DNS entries on your router or even on your PC to get the browser to work right.

Step 6: Add the PCI Adapter and W-NIC to a Desktop PC (optional)

The process of installing a W-NIC in a desktop PC is identical to the installation in a laptop, with the addition of installing the PCI adapter. The PCI adapter is required to provide a standard PC (PCMCIA) slot for the W-NIC. Most of the W-NIC manufacturers supply these adapters in the PCI-bus form factor. The PCI slot is currently the most common type of expansion slot for desktop computers, which are rarely equipped with the PC slot.

To install the PCI adapter, you should perform the following tasks:

1. Shut down your computer from Windows (Start/Shutdown).
2. Turn off the power switch (this may be on the back of the case).
3. Remove the power cord.
4. Remove the case side panel (if a tower, the left-side panel).
5. Turn the case on its side.
6. Locate a convenient, unused PCI slot, and remove the panel.
7. Plug the PCI adapter securely into the chosen slot; secure with screw.
8. Reassemble the case, reattach the power cord, and reboot.
9. Install the PCI adapter's driver software when prompted; reboot.
10. Install the W-NIC by following the procedures in steps 1–3.

Remember to observe static electricity precautions when you touch the PCI card or the inside of the PC. If you have another style of case, follow the disassembly instructions from your manufacturer. If you want to manage your AP from the desktop PC, install that software as well. You must ensure that your SSID (network name) and encryption match those on your successful laptop installation. You can change both of these in step 7, but this will help you have an easy initial installation.

You probably need to use the optional SOHO router because most DSL/cable service providers only allow one simultaneous login, which would be taken by your laptop. The router allows you to add two or more computers to the Internet connection, without having to pay the ISP for extra addresses.

Step 7: Configure Security on the AP and W-NIC

We have left this step until last, because security settings on the WLAN will stop your communication like a brick wall, if they are not properly configured. If everything is working properly before you enable the encryption and other security features on the AP and W-NIC, you can be more certain where to look for problems, if a security change suddenly cuts you off from the Internet.

WLAN security has three basic levels. First, the AP authenticates a W-NIC that wants to connect through its SSID, or network name. The SSID is a 32-character value that must be set on any device that can connect to a particular AP. Second, IEEE 802.11b networks support a type of encryption called wired equivalent privacy (WEP), which uses a 40-bit key. Alternatively, 64- and 128-bit key algorithms are available on some systems. Third, you may run a VPN over a wireless link, as with any wired link.

We will not discuss security in detail, as it is covered in Chapter 6. However, our basic recommendation is to use all the security you can afford. You must realize that there may be some performance consequences to turning security features on. That is, the better security algorithms may decrease throughput on the wireless link. The only exception is the network name feature. In most cases, the lowest level of WEP, 40-bits, adds little overhead. However, you may notice performance differences when you use 64- or 128-bit encryption, or when you use a VPN. If you were using a modem at 56 kbps, there would be very little difference, but the IEEE 802.11 network is capable of running speeds of 1 to 22 Mbps (and to 54 Mbps soon), so the time needed to process the more complicated algorithms can have a significant effect on throughput.

Project: Simple Wireless Home Network

At the minimum, you should change your network name from the default (usually the manufacturer's name). As a matter of fact, this "feature" is probably responsible for at least some apparent incompatibilities between manufacturers. If your AP and W-NIC are from different manufacturers, they will never link up until they share the same SSID. These IDs are case sensitive, and some punctuation characters may not be supported, so keep track of what you enter for the value. Likewise, the pass phrase used to generate the shared keys for WEP encryption is case sensitive. For more information on these and other security topics, consult Chapter 6.

Project Conclusion

At the conclusion of this project, you should have a simple wireless network for your home or small office. You will be able to access the Internet from your wireless laptop or desktop computer. For the details on turning this into a full-fledged SOHO network, with both wired and wireless components, see the project in Chapter 4.

CHAPTER 4

Home Office and Small Office WLANs

Chapter Highlights

- Dual Wired and WLAN
- DSL and Cable Modems
- Router Features
- DHCP and NAT
- Choosing a Hub/Switch
- Backup Power, UPS
- Project: Wireless Home Office Network

Many practical applications for small office/home office (SOHO) WLANs supplement an existing wired network. In fact, the WLAN will undoubtedly be much more useful if it has access to several local resources, such as file servers, shares (shared directories in a network server), printers, and fax printers installed on other computers. In a normal small network, Internet access is made through a DSL modem or a cable modem. When a WLAN is added to the mix, it too can have access to those resources, if properly configured. That is why it is important to understand how wireless and wired resources can work together for a total solution.

Much has been written about the SOHO network environment. The term SOHO is a little overused, probably because it is so useful. Most networking people would describe the SOHO environment as one with a medium to low performance requirement. In comparison to a large network (which would usually occupy a large office), this is probably valid. However, in this day and time, we routinely expect performance on our Internet connection that would have been quite incredible on many large network Internet connections only five years ago.

For example, in the past, many companies seeking Internet access installed ISDN routers, with appropriate subscriber lines and ISP connections. A B-channel ISDN link is only 64 kbps, but it connects in under 5 seconds, so from that standpoint alone it is far superior to an analog modem, which takes 30 to 40 seconds to connect. However, these same companies are now insisting on so-called broadband access, which can mean a full T1 or ISDN-PRI line (the equivalent of 23 B-channels plus one D-channel). This type of connection yields a symmetrical 1.5 Mbps to the Internet.

In contrast, our "home" DSL and cable modem connections yield a typical 800 kbps downstream, although the upstream speed is often limited to about 128 kbps. In most cases, these minimums are truly *minimum* because downstream rates approaching 5 Mbps are sometimes achieved.[1] If your service provider is equipped for high-bandwidth connectivity, your distance from the central office (CO) or the fiber interface point is optimum, and they actually have lots of bandwidth from all the neighborhoods all the way to the Internet backbone (big "pipes"), you can actually achieve incredible speeds like this over your "home" or "small" office Internet connection. It sounds like this connection is not so small after all! For the most part, the only difference in home/small/medium-sized offices is the number of computers, networked printers,

[1]See the Bytes to Bits Tech Tip in this section.

routers, and hubs. All of these components are present in greater abundance in the large network. Only the very largest enterprise networks will have fundamental differences from their smaller brethren.

In this chapter, we discuss the networking and computing components of SOHO networks. Because this is a book about building your own WLAN, we also feature the addition of wireless networking components to the standard SOHO LAN. We also cover DSL/cable modem differences, simple NAT routers, and special considerations for your networking components.

Tech Tip: Bytes to Bits

Computer jargon can be intimidating and confusing, as illustrated by the use of the term *bytes* in everything from data storage to modem throughput. Here are a couple tips to demystify the bytes and the bits.

First of all, in advertising, bigger is better. So the confusion between bits and bytes is sometimes used to the advantage of the manufacturer or service provider. Here is tip 1:

$$1 \text{ byte} = 8 \text{ bits}$$

...and 1 kilobyte = 8 kilobits...and 1 megabyte = 8 megabits.

This was not done to confuse everyone, because it turns out that the smallest number of bits that a computer can use is 8. A bit is a binary digit with a value of 1 or 0. Everyone knows that computers use binary arithmetic to function, right? Well, binary math is based on powers of 2, so 8 bits is $2 \times 2 \times 2$, or 2^3. The early computer designers, who were also mathematicians for the most part, came up with a name for the block of 8 bits, the *byte*. The 8-bit byte is also very useful for encoding the letters of the alphabet and the ordinal numbers, although only 7 bits are actually needed. You can contain two *hexadecimal* digits within the 8 bits, another convenient computer design feature. Originally, some computers used an 8-bit instruction word, or a multiple of 8 bits. Most of our modern PCs use 32-bit words, and the newest use 64-bit words. The computer math and the hardware are simpler with byte-sized data chunks.

High-tech runs on abbreviations and acronyms, and we abbreviate kilobits as *kb* and kilobytes as *kB*, just for sanity. Thus, a 40-megabyte drive is 40 MB, but 56 kilobits per second is 56 kbps. So, okay, it is really handy to have bytes, as long as we know what they are. Time for tip 2:

> RAM memory is given in bytes, kilobytes, or megabytes
> Disk capacity is given in bytes, kilobytes, or megabytes
> File size is given in bytes, kilobytes, or megabytes
> Communication throughput is given in bits, kilobits, or megabits per second

That means that you will buy a computer with 128 MB of RAM capacity and a 100-GB hard disk. Then you will send a 160-kB file over your 800 kbps cable modem (we'll receive it at that data rate, anyway). It turns out that when data is sent over an analog modem, it is actually sent 1 byte at a time, and a couple of bits are added for synchronization and integrity. This means that 8 bits of data plus 2 bits of overhead equal 10 bits in each transmitted character. Synchronous traffic is a little more efficient than this because the overhead bits are not needed for every byte of data. However, in an Ethernet environment, many data bytes (from 64 to about 1500) are combined into a packet with source and destination addresses. For the Internet, our favorite protocol, TCP/IP, adds some more overhead, and the wide-area protocols (such as PPP) used by the DSL or cable modems add even more overhead. When all is said and done, we probably get the equivalent of our 2 bits back (don't you want to put your 2 bits in about now?). Here is tip 3:

bits per second / 10 ~ bytes per second

Wow, that simplifies our life! If you need to download that 5 MByte file, it is going to take roughly the same amount of time to transmit $5 \times 10 = 50$ Mbits. Let's see, if you have a high-speed line that can achieve 800 kilobits per second (0.8 Mbps), transmission will take $50/0.8 = 62.5$ seconds, or about 1 minute. That is about right, if you have ever downloaded a 5-MB file from a very fast server over your cable modem or DSL modem. Isn't it cool how math works out! Here is tip 4:

> 5-MB file → ~60-second download (fast server)
> 1-MB file → ~10-second download (fast server)

Practically speaking, we rarely get a really fast server, so it will almost always be somewhat longer than that, especially if we are downloading from a popular site (such as virus files during a scare). You also can go the other way; so tip 5:

bytes per second × 10 ~ bits per second

Thus, if you have a file that your browser says is downloading at 70 kBps, this is equivalent to ~700 kilobits per second (70 × 10 = 700). So now you know the rest of the story.

Dual Wired/Wireless Networks

Once you have added a second computer to your home or small office, you will begin to need the benefits of a small network. Let us assume that you have two computers, one primarily for your own use and one for the rest of the family. Attached to your computer, you have a relatively high-speed laser printer, a combo scanner/printer/fax, and an access modem for DSL or cable. You have just added (or perhaps retired your old PC as) a computer in the family room or kitchen for your spouse and the kids, as well as your own convenience. This second computer has a color ink-jet printer attached.

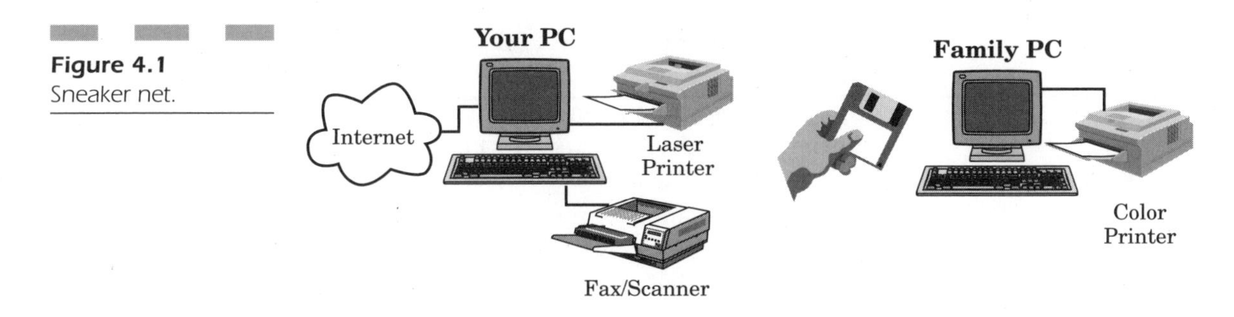

Figure 4.1 Sneaker net.

You now need the family computer to have access to the Internet and your higher-speed printer. Occasionally, when working at your primary computer, you need to print color documents, perhaps for reports or a presentation. The family may need to scan a document or send out a fax using the peripherals attached to your computer. Also, you would like to have access to your files from either computer, and you need to back up to the CD-RW writer (or DVD writer) on your PC. You could resort to "sneaker-net" (Figure 4.1), which is the unofficial term for carrying a data disk between the two computers (you don't really have to run or wear sneakers), but that really doesn't work well for the printer, fax, or

Internet connections. They really need to have a continuous data connection to the computer that is using them.

What you need is a small, home network. Although large offices with hundreds of computers have arrays of servers (and technicians to care for and feed them) running complicated network operating systems, such as NT and its derivatives, you do not need anything nearly so complex.

Peer-to-Peer Networking

Fortunately, Windows, Linux, and many other OSs provide for a very simple ad hoc network called *peer-to-peer networking*. In a conventional computer network, one or more servers support many PCs. The servers and the PCs are very different, both in software and function. The server is a specialized PC running server software, whose functions are to authenticate PC users, interconnect PCs to the server and each other, and to store centralized data files. In addition, the server may provide a point of connection to common resources, such as printers, routers, and backup devices.

This architecture is called *client/server* (Figure 4.2). The individually networked PCs act as clients, rather than peers[2] of the server. In Windows NT (and related OSs), each PC logs in (attaches logically, once the proper ID and password have been provided) to a *domain* on the server (very different from a Web domain on the Internet). Essentially, the PCs cannot "network" without the server because all of their communication is to and through the server.

However, there is another method to connect PCs into a network. Two or more PCs can form a peer-to-peer network. In Windows, this peer-to-peer grouping is called a *workgroup*. With the proper workgroup ID (called the workgroup name), any PC that has physical or logical connectivity can join the peer-to-peer network. Keep this fact in mind because it will become important when we discuss network security. In some cases, direct physical connectivity is not needed because the NetBIOS protocol that Windows uses to form the workgroups can be encapsulated within TCP/IP and transported across routers to locations far, far away.

Most simple SOHO networks will use peer-to-peer networking because it is very easy to implement, requires no special software, and provides a measure of security. Peer networks provide almost all of the

---

[2]A peer is a network component of equal status. Thus, the PC clients are peers of each other but not of the server.

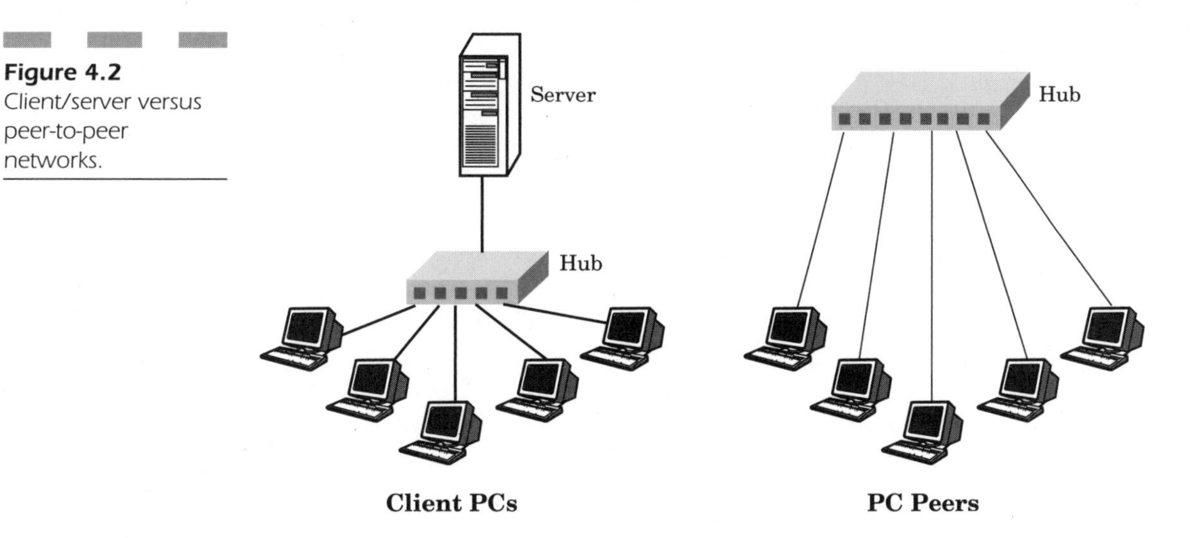

Figure 4.2 Client/server versus peer-to-peer networks.

resource-sharing advantages of more robust client/server networks, with a small fraction of the additional administration, hardware, and costs. From the point of view of a WLAN, there is little difference between a peer and a client/server network. Basically, whatever you had to do to connect a wired PC to the peering network, you must duplicate for the wireless PC. True, you also have to configure the W-NIC and the AP, but the rest of the process is just pure (or rather "peer") networking, the same as with wired PCs.

The project in this chapter shows you how to create a peer-to-peer network and supplement it with WLAN connectivity.

Centralized Resources

What sort of centralized resources might you want to have in your SOHO network? A centralized resource would be any resource that you would want to share between two or more computers. In the previous example, we had two different types of printers, one high-speed and one color, plus a fax machine and an Internet connection between the two PCs. Even though you might be able to tolerate whatever printer was connected to each PC and would put up with the inconvenience, there would be no way to get the Internet connection from your computer to the family computer. That would require a second line, at considerable expense, or you would have to share your computer with any family member who wants to get on the Internet. If you have children, between them and their friends, that could totally cramp your style, not to mention the fact that you have absolutely no room for the 1000 or so MP3 music files they would want to download and play on your computer.

Many families have recognized that the Internet is becoming a critical resource that must be available to more than one family member at a time. If you don't want to spend all those late nights waiting for your turn on the Internet, you should make plans to get Internet everywhere in your house, or at least on all the computers in your house.

In addition, it is very convenient to have access to a fax machine or a centralized file storage or backup unit. Also, you can easily and quickly transfer files across the network. This is great for backing up your data, which everyone needs to do on a regular basis. By using a common disk for your backups, you can periodically transfer all your documents and data files from the other networked PCs and use a CD-writer, DVD-writer, or tape unit to archive the files. You can even set up an automatic routine to do this backup at a particular time of day, each and every day. In Windows, this feature is called Task Scheduler, and it is available in the System Tray in most Windows 98- and-up machines.[3] The system tray is the rectangle filled with several small icons on the lower right portion of the desktop, in the Task Bar, right next to the digital clock.

Add the Internet

To add Internet access to all the computers (wired or wireless) in your SOHO network, you will need to add a hub. You do not need to implement peer-to-peer (or client/server) networking to get access to the Internet, only to share resources connected directly to other computers in your SOHO network. We will talk about choosing an Ethernet hub or switch in a later section, but for the time being, you just need to understand that the multiport hub is a component that is necessary for more than two devices to interconnect. In our SOHO example (in Figure 4.1), we had two un-networked computers, one of which had a direct connection to an Internet-access device. We wanted the other computer to have access to the Internet as well.

In Figure 4.3, you can see how we use the hub to give both computers access to the Internet. Our Internet connection is a high-speed access device, such as a cable or DSL modem, with or without routing functions

[3]Access to the Task Scheduler also can be made from Start/Programs/System Tools/Scheduled Tasks.

(described later in this chapter). The hub functions like a sophisticated party line, in that each computer can have an individual Internet conversation, virtually simultaneously. The hub effectively shares the Internet access mutually between the two (or more) computers.

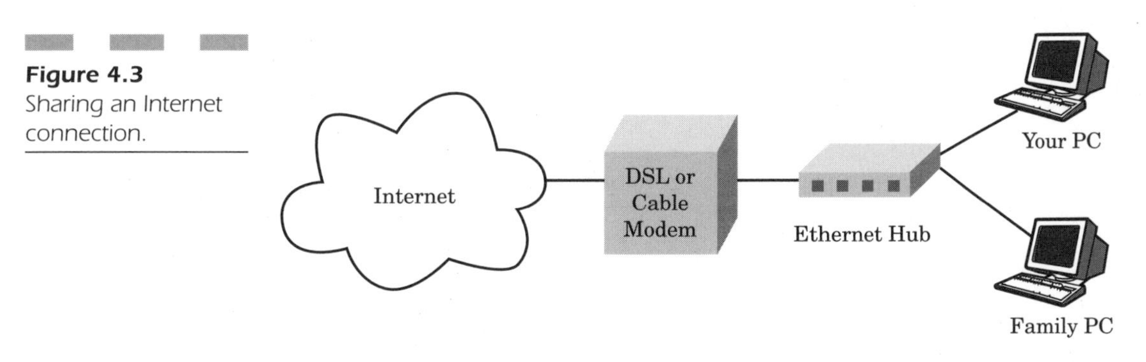

Figure 4.3 Sharing an Internet connection.

One of the marvelous things about Internet protocol,[4] abbreviated as IP, is that it contains its own routing information. Each IP packet contains the unique binary "address" of the destination device and the sender's own binary address, called the *source address*. The network devices that form the Internet itself are able to use this destination address to carry the packet to the proper computer or server elsewhere on the Internet. Once the packet is received, the receiving server simply looks at the sender's address contained within the packet, adds the reply data, and constructs a new packet with the originating address as the new destination and with the server's own address as the new source address. This return packet is directed back to the original computer to complete the miniconversation.

When you go to a Web page, through a series of send–receive actions, your web browser requests the page content from the server, one item at a time. A typical Web page may contain more than 50 such content requests, which must be fetched before the page is complete. Text and formatting for the page typically consist of a very small amount of data, whereas pictures and graphics may be 1 to 50 kB long. More graphics equals more time to load a page, so most good web page designers plan

[4]IP is a part of a group of protocols, which we call a *suite,* called Terminal Control Protocol/Internet Protocol, or TCP/IP. TCP/IP defines a series of small transaction-oriented communication snippets that are used for control, information, and communication. The structure of the packet is basically the same; only the details change and are extended to create the other protocols in the suite.

for a page to run no more than 30 to 50 kB, so even modem users do not have to wait forever. We high-speed access folks experience almost instantaneous *Web page painting*, as it is called.

A feature of Ethernet is that the receiver of your Ethernet card will only pay attention to packets that contain its unique address as the destination address.[5] Thus, your computer will "see" (or "listen to") only the response packets that are directed to it; conversely, the family computer will only see only its own Internet response packets. Though all the packets are echoed to each and every device connected to the Ethernet hub, each packet is selectively "listened to" by the appropriate device. Each packet that your "family" computer sends to the Internet is listened to by the DSL/cable modem, which sends it out to the Internet. The packet is also sent to your main computer (and any other that is connected to the hub) at the same time, but your computer ignores the packet because it does not have your computer's address as its destination.

This would be so cool if you could do this at the dinner table, or the office, or in a crowd at the station (or airport). Imagine that you could have a perfect conversation with one other person, no matter what the noise, distractions, or other conversations. Well, Ethernet and TCP/IP have accomplished this for the net. All we need to do is get ourselves wired (with the appropriate WLAN technology, of course), and we could all communicate perfectly. On the other hand, that may be what our teenagers are already doing, as they seem perfectly capable of ignoring our conversation, even when it is supposedly directed at them.

Wireless to Wired

The wireless AP is the device that connects the SOHO wired network to the wireless devices. The formal term for this internetwork interface is *bridge*. We say that the AP bridges wireless and wired devices. In a general sense, a bridge is a network device that repeats the signal between two Ethernet networks.[6] The AP takes whatever packet is transmitted by the wireless PC through its W-NIC and converts it to a packet on the wired portion of the network. Likewise, the AP takes packets destined

[5]The actual behavior of the Ethernet interface, coupled with a higher-level protocol, such as TCP/IP, is a little more complex than this. In addition to the logical IP addresses, each packet also contains a media access control (MAC) address pair, which is unique to the physical hardware send/receive pairs all along the path the packet takes from origination to ultimate destination. If you are interested, there are a number of excellent books on Ethernet and TCP/IP that go into the details further than we need to here.

for the wireless devices, captures them on the wired LAN side, and converts them to wireless packets to the destination PC. This is illustrated in Figure 4.4.

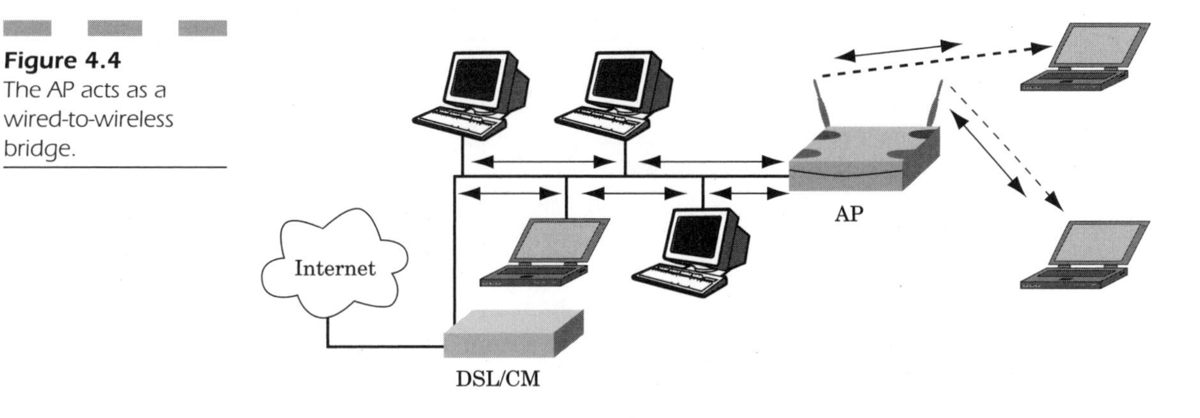

Figure 4.4
The AP acts as a wired-to-wireless bridge.

In operation, a bridge is not simply a repeater of packets. As you can imagine, in our WLAN example, this might present a problem. For example, if we had a large amount of traffic on the wired LAN (a large file transfer, for example), and that traffic was taking up a 10- or 20-Mbps portion of our wired LAN (assuming a 100-Mbps LAN), that would overload the lower-bandwidth WLAN. Remember, we said that a typical 11 Mbps WLAN could achieve only about 5- to 6-Mbps throughput in the real world, and that 802.11 automatically drops from 11 to 5.5, to 2, and to 1 Mbps when the signal is not optimum. Thus, 20 Mbps just can't fit into a 6-Mbps path.

Fortunately, a bridge "learns" the MAC addresses of the devices on each side, and it only passes those packets destined for the other side. The bridge would listen at full "line" speed to the large file transfer in our example, but would not pass any of the packets to the wireless side. Of course, if the file transfer *were* to a wireless device, then the packets would be passed across the bridge. Another neat thing that the AP bridge can do, as with all bridges, is to restrain the data flow to a sustainable rate. That means that, if the current wireless connection can

---

[6]That is, between two *collision domains*, a 50-cent term for a portion of a complex network where simultaneous, interfering packet transmissions, called *collisions*, are possible. A bridge is an OSI Layer 2 device, which means it functions at the MAC level only and is transparent to higher-level protocols, like TCP/IP. Wireless "Ethernet," IEEE 802.11, is merely Ethernet-like, so the AP doesn't meet the rigorous definition of an Ethernet bridge, but it functions essentially the same.

support, say, 2 Mbps, the bridge will cause the sending device to back off to that rate or less, so no packets are lost.

In this manner, the wireless network can operate rather seamlessly with the wired network, almost as if they were part of the same physical entity. In a way, they are the same logical network. IP network address space is divided into sections (segments are another concept in Ethernet networking) called *subnets*. This subdivision defines a group of contiguous IP addresses, like 101, 102, 103, etc., that are all in the same subnet. To get to any address in your subnet, your packet is usually made available directly to that device through a hub or a switch. To get to a device on another subnet, you must go through a device called a *router*. Simply stated, the router has the ability to identify a route to a destination device by looking at the subnet in the packet and forwarding the packet along a path that will result in the packet being properly delivered.

The bottom line is that your WLAN devices (those with W-NICs) will need to be part of the same IP subnet as the wired LAN devices. You will need to set that up in the configuration of each PC you connect over a WLAN, just as you did for the wired PCs. If you do not do this, your wireless devices will not be able to connect to anything in your wired network, even the Internet router. The packets will be there, but they will just be ignored, through the selective hearing of IP.

Cable Modem and DSL Differences

Throughout this book, we assume that you have a high-speed Internet connection. This high-speed connection would be provided through a cable modem or a ditial subscriber loop (DSL) modem. There are some important differences between the two technologies, and we need to point them out. For the most part, both connections provide a much higher-bandwidth connection to the Internet than conventional (old-fashioned, perhaps) analog modems. That is, they can give you a connection that ranges from about 256 kbps to over 1 Mbps.

That is definitely high speed when compared with conventional modems. Regular analog modems, once called telephone modems, operate by converting digital data signals to analog tones. These tones are then carried by ordinary telephone technology. For their time, analog modems worked very well. The data rates that could be carried by analog modem technology increased from a crawling 300 bps to a much faster 56 kbps[7] rate over 20 to 25 years. This was done by more cleverly encoding the data, compressing it, and providing error correction. On many phone circuits, that was simply not enough, so most of us put up with one of the "fall-back" speeds of 33.6 kbps or even less.

The problem with conventional telephone line transmission is that the data has to be handled within an audio (voice-range) bandwidth of only 4 kHz. After all, the phone system was designed to transport voice, not data. This is really bad news in the days of the World Wide Web. For the first time, we have a routine need for much higher data rates than needed for simple character transmission. You see, a normal computer screen in the text-only days was comprised of only about 2000 characters, each 1 byte wide. That is only 2 kB, which doesn't take very long to transmit, even at 2400 bps (~8.3 seconds for you math whizzes), but the Web added pictures and graphics. That multiplied the page download by about a factor of 25 or more. Now, at the same 2400-bps rate, that Web page would take nearly 3.5 *minutes*, clearly unacceptable. Even at 24.4 kbps, you would have to wait 30 seconds. That is an interminable delay for most of us.

The answer to this dilemma is to find another technology to get data to the browser on your PC. The development of cable modem technology and the adaptation of breakthroughs in digital subscriber line technology offer the ability to multiply data rates by a factor of 10 or more, usually much more. Most homes and businesses in the United States and Europe have access to DSL or cable modem lines. For more on the details of DSL and cable modem operation, refer to the Tech Tip on How DSL Modems Work and How Cable Modems Work.

Tech Tip: How DSL Modems Work

DSL modems bring high-speed Internet to your home or office over an ordinary two-wire (one pair) telephone line. The way they do this trick is really fairly simple, although it takes a marvelous combination of modern data and signal processing technology to work the magic. Basically, the DSL modem piggybacks a high-level digital signal on the phone line and uses simple but effective filters to keep voice and DSL signals apart.

[7] As everybody knows, FCC restrictions do not allow 56-kbps modems to run any faster than 53.3 kbps, but do we really care? Nothing really works to its full specs, it seems, but advertising.

Not everybody knows that the ordinary telephone line supports two-way transmission of voice frequencies from about 100 Hz to around 4,000 Hz. This nicely encompasses all of the audio frequencies needed for good voice fidelity. Amazingly, this simple, twisted pair of wires can carry your voice over 4 miles back to the telephone company's central office (CO). In practice, phone companies try to limit the cable distance from CO to subscriber to a little over 2 cable miles, specifically less than 12,000 feet. More than 85 percent of telephone subscribers in the United States are within this distance from their COs.

Telephone engineers have known for some time that the local "cable plant," as it is called, could support frequencies far above 4000 Hz. As a matter of fact, over the past 30 years, a technology called T-Carrier has carried 1.544 Mbps data signals over two cable pairs by using extended frequencies. These classic T1 lines operate at a nominal 6000 cable feet and can be extended by the use of in-line repeaters to almost any distance. However, T1 does not coexist with voice frequencies on the same line, and it requires two pairs rather than one.

About 10 years ago, a technology began to emerge that allowed specialized transceivers to send data on a single cable pair by using frequencies above the 4-kHz voice range. To accommodate the different lengths of circuits from CO to subscriber, this technology would decrease the data rate as the cable distance (and attendant losses) increased. In addition, because it did not use the voice frequencies at all, it could be placed on a phone line that was already in service for voice traffic, such as the regular dial-up phone line in our homes and offices. This technology was called "digital subscriber loop" because it allowed the copper pair to the customer (called the subscriber loop) to be used for digital data. It probably should have been called an analog–digital subscriber loop, but DSL is more inclusive of all DSL technologies.

One of the first implementations of DSL technology allowed the up- and downstream data rates to be different, or asymmetrical. Naturally, this became ADSL, which is the form of DSL that is used in more than 95 percent of the installations to date. Because (relatively) old-fashioned ISDN is also a form of digital loop, it has picked up the moniker of "IDSL" on occasions, but it really is not in the same ballpark, because it absolutely cannot coexist with plain old telephone service (POTS) on an analog phone line as ADSL can. There is also an SDSL, or symmetrical DSL, and a few other variations, such as HDSL, or high-speed DSL, which provides T1 or E1 rates over a single pair, using similar adaptive technology. Many started calling these technologies xDSL or XDSL, which was just plain confusing, and you know the marketing folks at the phone company don't like confusion, so everybody uses "DSL" for their ADSL service.

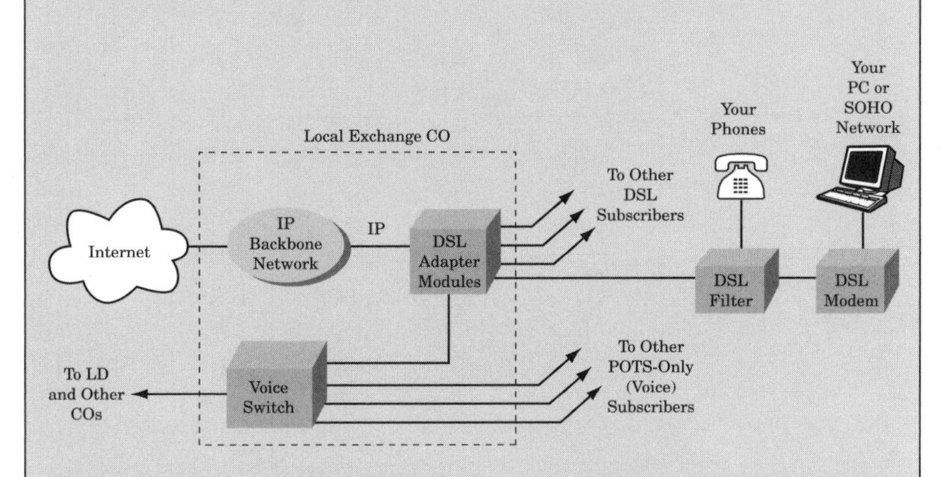

The speed of a DSL line can range from 1.544 to 6.1 Mbps. Upstream and downstream speeds vary according to line conditions. Your provider probably uses speed limits that provide a margin of safety and may artificially limit the speed to allow for cost-based bandwidth rates and prevent you from overloading their network. You could purchase your own DSL modem and install it yourself. However, there are several implementations by different manufacturers, and you probably will need to obtain your DSL modem from your local access provider. Also, at this time, most providers give you a DSL modem as part of their service, so there is no real benefit in buying your own modem.

Tech Tip: How Cable Modems Work

Cable modems use the CATV system to bring you Internet access with your cable television service. The way this works is fairly simple. An unused portion of the cable system bandwidth is used to send IP data to your home over the same cable that brings your TV signals. Unlike the telephone system that was designed for the narrow 4-kHz bandwidth of voice, your cable TV system is good up to about 800 MHz. Most of that bandwidth is occupied by 6-MHz TV channels plus a portion that is used for FM music channels. The upper range of frequencies is often used for new digital TV channels and the new movie-on-demand service.

Cable modem technology sends data *downstream* to your home on one of the unused upper channels. Your cable modem sends data back *upstream* to the cable provider on one of the unused lower frequencies at 5 to 42 MHz. This lower frequency range is below TV Channel 2 (at 54 to 60 MHz) and theoretically is the best (lowest loss) frequency range on the cable.

A cable modem's upstream and downstream data rates are asymmetrical. This is appropriate because the primary use for cable modems is for surfing the Web and e-mail (one might suggest, for receiving junk mail). Most modern cable modem systems use the DOCSIS (data over cable service interface specification) standard. For downstream operation, these modems can use any 6-MHz channel from 88 to 860 MHz at a physical (encoded) speed of 34 or 43 Mbps with error correction. On the upstream side, the modem users can send data at 200-, 400-, 800-, 1600-, or 3200-kHz bandwidth increments with a frequency range of 5 to 42 MHz.

The downstream side is normally limited by the bandwidth of the fiber optic connection back to the cable provider's backbone and then by their bandwidth "pipe" into the Internet backbone. However, the upstream bandwidth is configurable by the provider. In this way, they can charge businesses more for greater upstream bandwidth. The average home user has little need for high upstream bandwidth, so 200 to 400 kbps is fine.

Cable modem technology requires that the cable operate two-ways. That is, amplifiers, taps, distribution cables, and drop cables must be able to support at least a limited reverse-channel capability. A method for getting the data signals all the way back to the cable provider is needed because each subscriber's cable is split off a series of distribution cables that go back to the cable "head end." It would not be feasible to try to send hundreds or thousands of user data signals back to the head end over a supersized reverse channel, so the cable TV industry has begun to install distribution cabinets, fed with both CATV and fiber optic cables, in local neighborhoods, to insert the downstream data channel, split off the upstream data from the subscribers, and transport that data over fiber optic cable to the cable operator's data center.

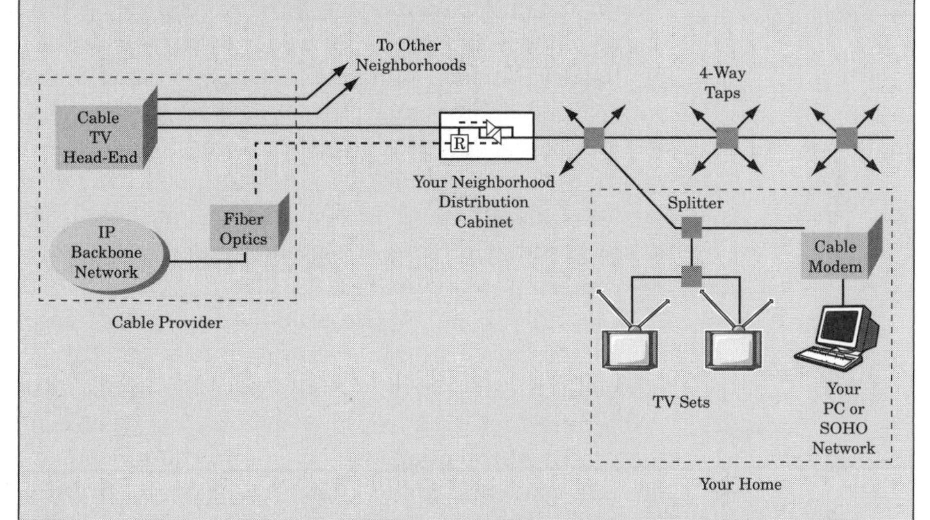

An important data security difference for cable modem technology exists because you, the user, may share a two-way cable segment with all your neighbors, some of whom also have cable modems. In some newer systems, the cable company is actually running individual coax or fiber optic lines from each home back to the neighborhood distribution cabinet. However, until that is done, you will be rather exposed to intentional snooping, or stealth hacking, through your cable segment. Stealth hacking is when a computer virus deploys a remote client (called a Trojan Horse) on one of your neighbor's computers (or on your computer), and the hacker uses that remote client to snoop into the other computers on your cable segment. See the cable modem security topic in Chapter 6 on WLAN security for more information.

Throughput

One of the biggest concerns when comparing cable with DSL modems is the data throughput of each type of modem. Unlike standard telephone modems, which can operate at the highest speed within their capability, both types of high-speed access modems are usually artificially limited by the constraints of the communications line or the service provider. Let's talk about these constraints.

The first limit on a cable or DSL modem is the length and relative quality of the transmission line. As you can see from the Tech Tip, the cable modem uses a CATV line, usually a coaxial cable at the house, to transport the data signal to and from a concentrator, and then on to the cable head-end, or more often, to a fiber/cable concentration point. The DSL modem operates over "ordinary" telephone lines, often the very phone line you use for your home phone. Out of the box, both modems are technically capable of providing as much as 4 to 6 Mbps of data throughput. However, practical considerations of the cable make this top-end rate very unusual.

For cable modems, the practical considerations are the quality of the cable line from the data concentrator (somewhere in your neighborhood), through all the in-line signal amplifiers and splitters, right down to the cable line that goes from the pole (or splice box, if buried cable) to your house. Even the quality of cable and connections within your house have an effect on this signal. One of the biggest culprits in cable signal degradation is signal "ingress," that is, leakage from the outside world directly into the cable. And cable modems are very unforgiving, unlike most TV viewers; interference just can't be tolerated, whether it is from poor signal strength, signal leakage, or just noise. To compensate, most cable providers plan for a lower data throughput speed, which can tolerate poorer transmission parameters.

With the DSL technology, special equipment at the telephone company's CO places the digital signal onto the phone-line pair; conversely, the DSL modem at the home (the subscriber) captures that digital signal for use by your PC. As described in the Tech Tip, an ordinary analog phone service can co-exist with DSL. However, DSL is much more sensitive to line parameters, such as length, capacitance, and noise sources. Although a telephone voice signal can easily go 20,000 feet from a CO to a house, the DSL signal begins to degrade rapidly before finishing that distance. For optimum performance, you should be within about 6,000 feet of the CO. Operation is still practical up to 12,000 feet, and special equipment may be required to go up to 18,000 feet. As the distances are extended, the DSL modem will adjust its signal, and the resulting data throughput will decrease. Some installations may take the DSL link down to ISDN speeds, as low as 128 kbps. Not bad, compared with an analog modem but still way under expected performance.

The second way in which DSL and cable modem speeds may be limited is intentionally, by the carrier (the ISP). Think of the ISP's data pipe to the Internet backbone as an expressway (Figure 4.5). Traffic can go real fast, but only so much traffic can be accommodated before congestion occurs. You have a really fast car, or maybe a whole fleet of cars, and if you are allowed to run at whatever speed you want, with as many of your fleet of cars as you want, you will cause problems for the others who want to use that same expressway. The ISP can limit your speed into its network so your use doesn't overload its internal data network's connection to the Internet backbone. This is a phenomenon that exists for operators of cable modem and DSL systems, so don't be fooled by raw claims about link speed.

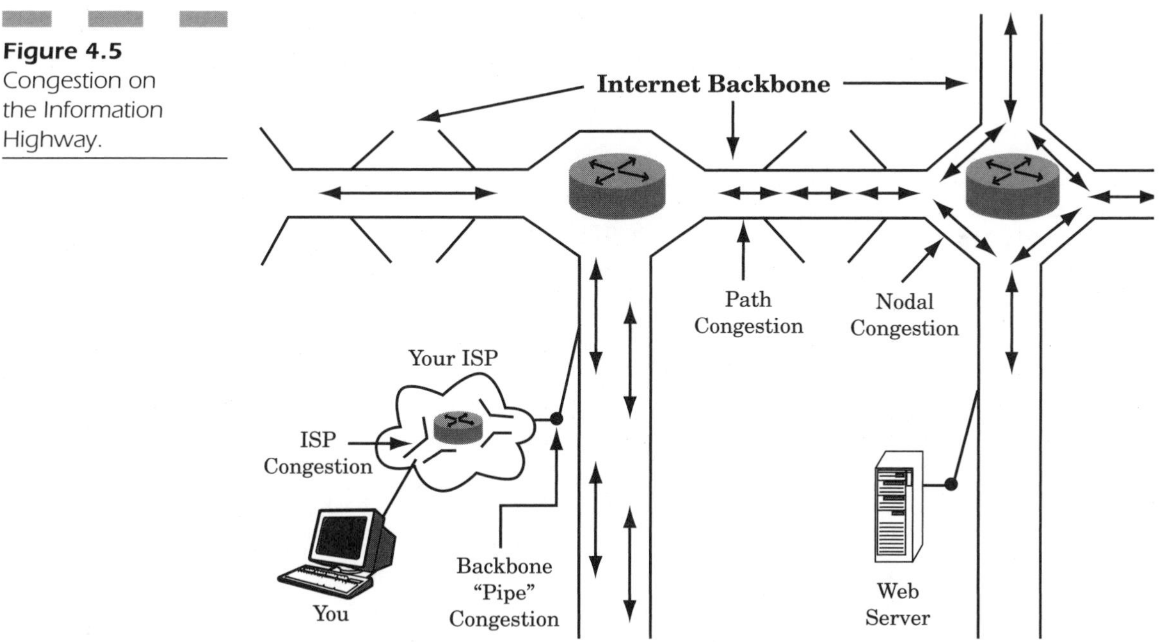

Figure 4.5 Congestion on the Information Highway.

Throughput is the actual rate at which data can be sent or received to/from a Web server on the Internet. Throughput can change from moment to moment as conditions that produce congestion (and thereby reduce responsiveness) change on the path between you and that particular web server. If you get less than optimum response, there is some sort of capacity restriction at some point along that path. For example, your local cable segment may have lots of data traffic; or all the other users, combined, may be exceeding the ISP's capacity into the backbone; or a particular leg of the path may be congested; or the particular server you are accessing may be jammed up. Think of what happened when a certain provider of "secret" undergarments did a Web cast of their spring fashion show. The resulting traffic severely taxed the Internet. In the same way, the servers handling game day statistics, an important news item, or severe weather may get clogged up. The maximum possible throughput would be obtained at a time of very little traffic along the entire path to a high-capacity, but lightly used, server.

Practical Considerations

Make no mistake about it, DSL and cable modem operators are in direct competition. You will see claims and counterclaims galore. The DSL providers will say in non-techie terms that a cable modem segment has a very limited bandwidth, that the very best speed you can hope for is when no one else is on, and that the bandwidth becomes so tiny as to be unusable, except in the middle of the night. The cable modem providers will counter that DSL has a much more limited speed than they do, and that DSL is not available in most neighborhoods. To an extent, both claims are partially true. You need to know to what extent, and to what significance, it is to you.

Let's look at the bandwidth claims first. It is true that the "segment" of a cable modem neighborhood represents shared bandwidth. Simply put, the cable modem uses a 6-MHz slice of the cable bandwidth for upstream and downstream data. A typical "neighborhood" consists of 50 to 200 homes (or businesses). If a bunch of your neighbors are all doing high-speed data transfers (those nasty MP3s and such), your favorite Web site will load a little slower, but it certainly does not crawl. First of all, statistics show that fewer than 10 percent of homes with cable connections will opt for cable modems. That means that you are really only competing, on average, with 5 to 20 other homes. If the cable is capable of about 6 Mbps of throughput, you still get as little as 300 kbps and as much as the full 6 Mbps...not bad! Second, people usually paint Web pages and then take their time reading them. No one constantly loads Web pages. File transfers, such as MP3s, are the biggest complaint, and many cable providers have taken steps to limit the available bandwidth for such traffic. This is much like a really smart traffic light at the expressway on-ramp that lets ten cars through for every two trucks.

The DSL provider has a different bandwidth problem. It is really easy to overload its internal backbone and Internet pipe with excessive data traffic. However, it is also very easy for it to bandwidth limit its users because each user comes into the CO on an individual pair of wires. This means the provider can charge more if you want more bandwidth, on the basis that you are having a greater impact on its network and thus its costs. On the other hand, all most users really want is fast access a few times a day, but you will most likely be limited to some arbitrary data rate, as low as 128 kbps. Fortunately, most DSL providers give their low-cost users at least three times that, or 384 kbps. Some even provide 768 kbps of link bandwidth to more or less match most industry claims for cable or DSL of about 800 kbps. However, a few (probably by accident) still provide the 1.5 Mbps (or more) that the technology is capable of, but only if you meet the distance limits back to the CO. By the way, this distance is "cable feet," or the distance along the actual cable path to the CO. You may be able to throw a stone to the CO, but still be 6,000 cable feet from it, electrically.

Cable and DSL service providers take it on the chin when it comes to service availability. Neither the cable TV system nor the telephone cable system was designed to support data traffic, so it is a stretch for both. In each case, equipment has to be added, and in some cases, the base cable plant must be replaced. When you consider that the technology of the telephone network has basically remained the same for more than 100 years, it is not surprising that some of the cable installed 40 or 50 years ago would be inadequate for DSL. By the same token, the CATV cable, amplifiers, and splitters that were installed a mere 20 years ago were not designed for the number of channels that are in the extended cable TV offerings, much less the existence of a reverse (data) transmission.

From a practical standpoint, there are many places where one high-speed technology is available, and another not. There are also many neighborhoods that have both, and a few that have neither. The point is that your choices for a high-speed Internet access technology may be limited, and you need to understand the reasons for those limits. Also, you may want to consider alternative medium-speed Internet access, such as provided by ISDN, point-to-point wireless access, and even satellite links. Also, conventional T1 and fractional T1 service may be available, even where DSL is not. You may have to pay a few dollars more, but if your need for bandwidth is significant, relative to the costs, you should consider these alternatives.

Security Concerns

Another difference between classic DSL and cable modem connections is their relative security to the average user (Figure 4.6). We have a whole chapter on security, so we will not dwell on it here. However, you should understand that a Windows user with no special precautions might be somewhat more vulnerable on a cable modem segment than on a DSL line. The increased risk is slight, and highly dependent on whom you share your cable with, but it does exist.

In a nutshell, anyone on your cable segment can potentially access your hard drive and all your files through Windows networking. Remember that Windows networking (file and printer sharing) uses a very simple protocol called NetBIOS. If you are so unwise as to leave your default workgroup setting in place (the default is "Workgroup"), then virtually anyone on that cable segment can have free rein to your files. Even if you change to another workgroup name, you may be somewhat more susceptible to direct attacks. Probably, it is just some kid playing around, but it also might be a more serious intruder.

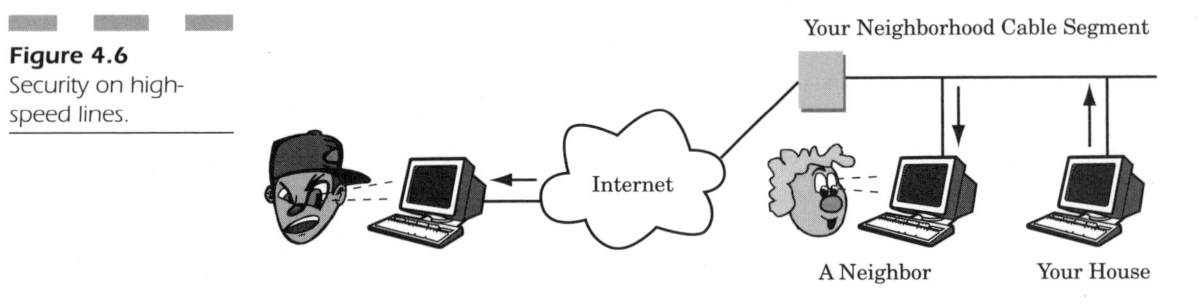

Figure 4.6 Security on high-speed lines.

DSL is no panacea. Although direct workgroup access is more difficult, it is possible, and there are a host of other access methods that can be easily identified. As a matter of fact, all those NetBIOS messages may be encapsulated within TCP/IP and accessible from anywhere.

There are a number of steps you can take to increase your network security. Among these are to change all default passwords (and remember the changes), use a NAT router, turn off as many network "holes" as you can, use a personal firewall, and keep your virus scanner up to date. Check out the next section for a more complete description of the advantages to a SOHO router. Chapter 6 on WLAN security covers these issues, in addition to wireless security.

Fiber Links in the Future

When it comes to throughput, fiber is king. It is quite possible to convey Gigabit Ethernet[8] over fiber directly to the home or office. As a matter of fact, virtually every city of any size is being wired with gigabit fiber, at least in the downtown and commercial areas. We suffer no fools when it comes to thinking that the DSL and cable modem bandwidths of today will serve us for all time. As with all things technical, its time has come, and its time will pass. Fiber links may be the logical answer. The concept is so powerful that it has inspired a brand new term: FTTC—Fiber to the Curb!

Here are the current practical constraints. Fiber is available in two basic types, single mode and multimode. The raw cost is only a little higher for single mode, but the optics for the adapter cards are at least three times higher. Multimode fiber can currently reach a bit farther than 500 m (about 1500 feet) for Gigabit Ethernet, whereas single-mode Gigabit Ethernet can range from 5000 to 70,000 m. Remember to think in cable feet because the fiber will have to twist and turn to get to our particular curb.

The beauty of fiber is that it is not limited to just one wavelength. The Gigabit Ethernet signal is carried on one wavelength of light, but you could add lots more to the same fiber with little or no interference. This is similar to different colors of light, although we use infrared light wavelengths for fiber optic cable transmission. Think of this: the fiber provider could put your phone line(s) on one wavelength, your Internet connection on another wavelength, and your television entertainment on another wavelength. This would only take three wavelengths and would be carried on the same fiber, right to your home or business. Practical *wave division multiplexing* technology for fiber optic cable can now put 4 to 120 wavelengths of light on the same fiber. Think of the possibilities!

Using a Router with Internet Access

One of the first issues you may run into with the typical DSL or cable modem high-speed Internet access lines is the fact that these devices really function as a "bridge" into the access provider's[9] network. A *bridge* interconnects two or more devices remotely[10] and is ignorant (transpar-

[8]For more information on Gigabit Ethernet and fiber optic cabling, please refer to *LAN Wiring* 2nd ed.

ent) of IP and IP addresses. However, the access provider typically assigns you only *one IP address*! That means you can physically connect only one computer or other device that answers to that IP address.

That is a major problem if you expect to have two or more computers connect to the Internet through your high-speed modem. The service provider does this to conserve IP addresses. Because the IP address space is not infinite (not by a long shot, although it was originally large beyond anyone's dreams), the users are essentially rationed valuable IP addresses. The normal arrangement with a service provider is for one address. If you want more than one IP address, you generally must pay for it. However, there is another way.

With a router (Figure 4.7), you can "share" your one IP address with several computers. If you look at Figure 4.3 for comparison, you see that a router has been added. In that example, the IP subnet of the local computers had to be the same as the IP subnet of the access provider (shown as the Internet cloud). However, that would have required that you get two IP addresses from the provider. In the early days, this was usually possible, and it still is with some providers. However, we are rapidly running out of available IP addresses (as we did with telephone area codes), and no more address space may be created without making our entire networking infrastructure essentially obsolete. Providers are now limiting their users to just one IP address, unless the users pay extra for more.

In Figure 4.7, you can see that your link to the Internet made over an IP subnet numbered, for the purpose of this example, 264.128.32.0[11] is the access provider's subnet that our high-speed modem is attached to. Of the 256 possible addresses available in this subnet, the provider has assigned 264.128.32.16 to us. The router uses that address to connect to the provider on our behalf. The side of the router connected to this provider subnet is called the WAN side and is "outside" our network.

---

[9]The *access provider* is distinguished from the ISP as the provider of physical connectivity to the user. In many cases, they are the same company (your local phone company or cable TV company), but you may have an access provider give you a connection to an outside ISP over DSL or cable modem lines. This is much more common with DSL than with cable modem service at this point.

[10]Technically, this describes a very simple wide-area network (WAN) bridge. There are also local bridges. A bridge theoretically can connect more than one device on either side, but the real issue with Internet access is the assignment of a static or a dynamic IP address.

[11]The pros will recognize that this address is totally bogus. Each of the four "octets" in an IP address must be in the range of 0–255. We made the first octet more than 255, so that someone's real address is not used. This is the same thing that TV and movies do with 555 phone numbers.

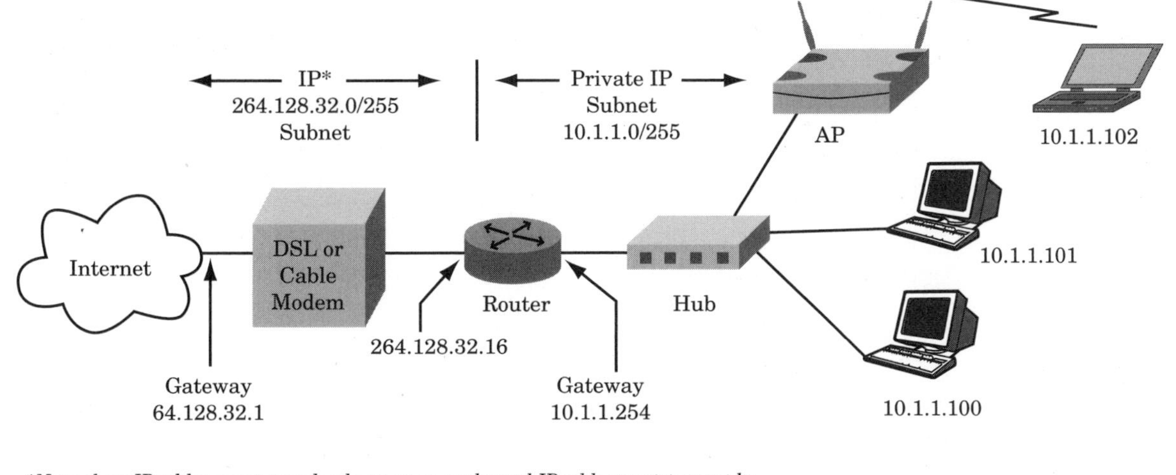

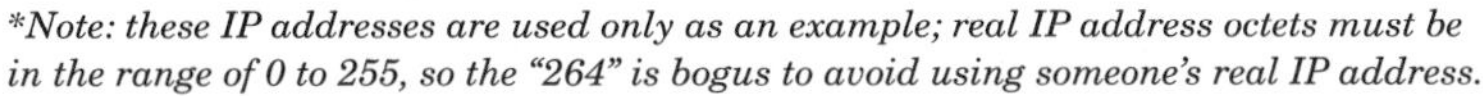
Note: these IP addresses are used only as an example; real IP address octets must be in the range of 0 to 255, so the “264” is bogus to avoid using someone’s real IP address.

Figure 4.7 Adding a router to a SOHO network.

The other side of the router is our internal, private IP subnet. We use a special category of “nonroutable” IP addresses for this private subnet. See the Tech Tip on Private Nonroutable IP Addresses for more detail. Simply put, there is a group of IP addresses that we can all use for our internal networks. Normally, the routers we use for sharing a DSL or cable modem line will automatically assign the addresses within the proper range for each of our computers. However, if you manually assign the addresses (you may need to assign a static address for a stubborn computer), you must ensure that you stay in the ranges specified in the Tech Tip.

Theoretically, you could use up to 254 IP addresses inside your private network (within each class C space, the fourth octet). However, you will need only a few. There are many advantages to this scheme of IP addressing. Not only can you easily add and remove computers and other network-addressable devices from your network without anyone’s permission, but you can also save money because you don’t have to pay for extra IP addresses for each of your computers. In addition, you now have IP addresses assigned for all the computers in your SOHO network.

It is easy to see how all the network traffic that is intended for your other local computers, printers, servers, and wireless components stays within your private network. The router notices that those packets are addressed to the internal network and ignores them. However, any

Tech Tip: Private Nonroutable IP Addresses

As the Internet expanded, it was necessary to accommodate large networks with many thousands of computers. IP address space was at a premium, and as companies grew their internal networks, each computer required a "real" Internet address to connect to the Internet. If an internal network was interconnected to the Internet, *all* of the internal computers had to have real Internet addresses. As addresses ran out, companies and providers began to grab up the available address space for their future needs.

The solution was to create three ranges of *nonroutable* IP addresses that could be used by absolutely any company. These nonroutable, or *private*, IP addresses are within defined ranges and automatically blocked by all the routers. Three ranges are defined as:

192.168.x.x
172.16.x.x through 172.32.x.x
10.x.x.x

Any address within these three subnets is free to use, without regard to conflict with a "real" routable IP address that belongs to someone else.

packets that are sent to the Internet will have an IP subnet address outside of your private network, so the router will send them across to the provider's gateway on the external network. The router easily recognizes these external addresses because they are of a different subnet.

Conversely, external Internet IP traffic that is destined for your private network will be sent to the external address of the router (in this example, 264.128.32.16), which will determine the internal address for which it is destined and resend the packet to that computer. Your computer doesn't know the difference, and neither does the network.

One word of caution: certain types of Internet communication may require a valid (real, public, routable) IP address. An example of this is the VPN, which is used to create a secure path back to your company's internal network. Special configuration of your router may be necessary to create a "pinhole" for a static IP address through the router. Such a pinhole makes the router act like a bridge for that one IP address, so that it does not convert the address to the internal subnet. This feature works well but may inadvertently create a security hole out of the pinhole. However, it may be the only way you can operate your VPN over your router. Some newer VPN technology can operate through a NAT device. See the section on NAT for more details.

DHCP—Dynamic IP Address Assignment

IP address assignment is one of the more frustrating aspects of IP networking. The operation of an IP network requires that your computer have an appropriate IP address on the subnet to which it is connected. Thus, if you are on the good old 264.128.32.x subnet, you must have an address with the same three-octet prefix, which does not conflict with any other address that is already assigned. Unfortunately, if you have a perfectly good, permanently assigned IP address, and someone else misconfigures their computer on that subnet to use the very same address, a conflict is created and everything will get very confused, assuming either computer can connect reliably.

A protocol called dynamic host control protocol (DHCP) was created to deal with this problem. DHCP allows for the "automatic" assignment of a valid IP address, within the proper subnet (Figure 4.8). This is usually done at startup (boot) time. Your computer simply uses the DHCP protocol to request an IP address from a DHCP server. The DHCP server gives your computer an open IP address, in return, along with other information, such as the gateway address and DNS. In this way, you can easily add computers to a network, with no fear that you would inadvertently pick an IP address that was in use.

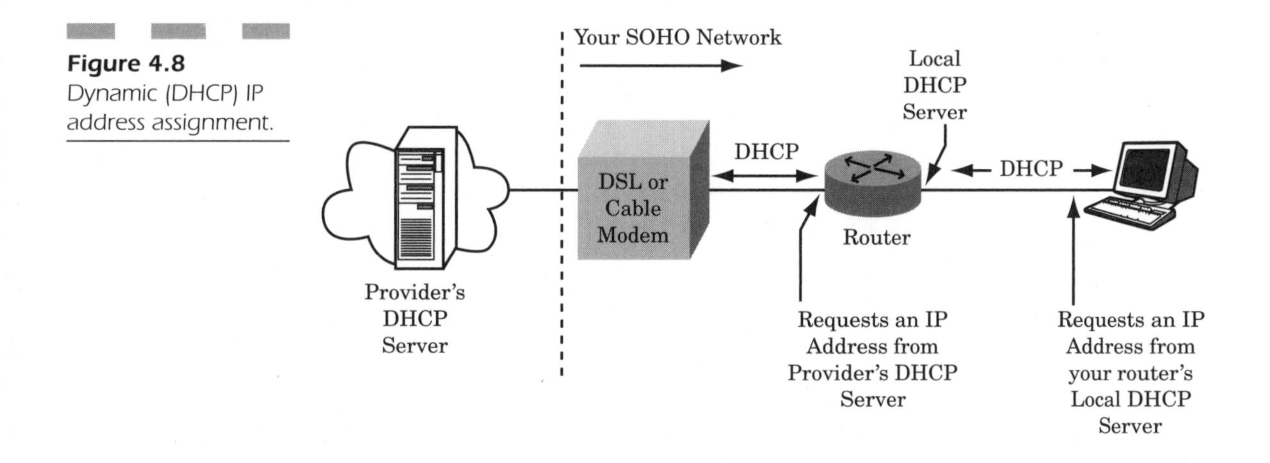

Figure 4.8 Dynamic (DHCP) IP address assignment.

In most Windows environments, DHCP (sometimes called dynamic IP) feeds all the appropriate parameters into the operating environment. You can take a look at these particular addresses by running ipconfig.exe in Win 9x or by looking at the properties for your network connection in NT operating systems (including XP). For more details on looking up these parameters, see the procedures in Chapter 3.

The alternative to dynamic IP address assignment is to manually assign an IP address to your computer. This manually assigned IP address is called a *static IP address* and will not change when the computer is rebooted. Most computers will not require a static IP address, other than the case of the VPN pinhole described in the previous section.

As described in Chapter 3, a problem with the network, internal or external to the router, may often be diagnosed by looking at the DHCP-assigned IP address. If the DHCP server is not there for some reason, the computer (or other device) may continue to boot, but will substitute a nonvalid IP address (nonvalid for the local subnet). Looking at the network parameters may show a 0.0.0.0 address, if DHCP failed the router. If you have a Windows device, the address will likely be 169.x.x.x (x is any address in the 0–255 range). If this happens, either you have a problem connecting to the network, something in your configuration is preventing it (e.g., you are not set up for DHCP to "automatically assign" an IP address), or the DHCP server is unreachable.

Network Address Translation (NAT)

Your SOHO router will allow you to have a private subnet for your small network. This small router can handle only a very simple routing function from your internal network to the rest of the world, but in doing so it also performs a network address translation (NAT). The type of NAT that your SOHO router does will allow several PCs on your local network to access the Internet through the single IP address provided by your service provider. This is illustrated in Figure 4.9. We won't spend a lot of time on this subject, because it is rather technical and there is a Tech Tip on NAT in this section.

To make your SOHO network work properly, you should understand that the SOHO router gets its IP address assigned by the DSL or cable provider using DHCP, and each of the PCs and other devices on your internal network get their respective IP addresses from the SOHO router. This includes the wireless AP. To make everything work properly, each device that is to assign an IP address will need to have that fea-

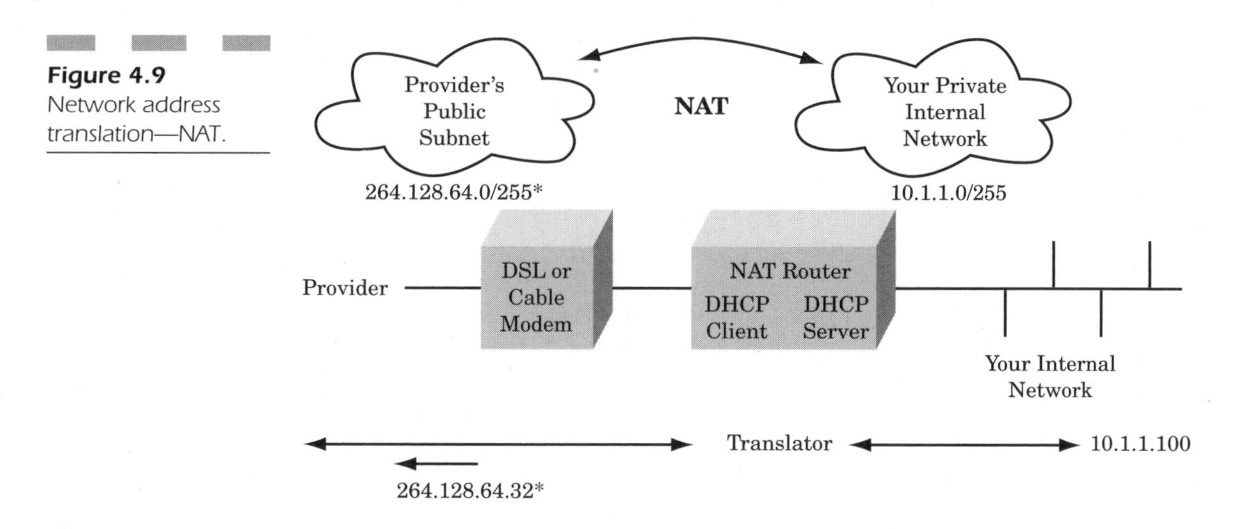

Figure 4.9 Network address translation—NAT.

ture (DHCP server) enabled, and each device that connects to a subnet likewise will need to be configured to automatically obtain an IP address (DHCP client). The DHCP request normally occurs on boot up, which may be at power on or when reset.

If you have occasion to turn off the power to any of the devices upstream from your PC, you may have to reinitiate the DHCP sequence. DHCP addresses are valid for a limited period and must be renewed. In some cases, the same address will be assigned for a renewal or a new request. However, in other cases, the address will be changed, and the old address will no longer be valid.

The correct power-up sequence is to start closest to the Internet. First, power off and back on the DSL or cable modem, then the router, then the PCs and AP on the local network. You will need to leave power off for at least 10 full seconds, and possibly 30, on each device. Wait for the device to go completely through its power-on routine (self test and to re-establish connections), which will be demonstrated by all the pretty flashing lights finally settling down to steady green (or, in some cases, off). Then go to the next device in sequence. You may not need to reboot the computers at all, which is at best a tedious process, so check them out with a quick attempt to reach your favorite Web site. If live content pops up, you are OK; but if you get a "server not found" or other error, you will have to reboot.

Tech Tip: Network Address Translation

In a typical company, at any given time, only a fraction of the computers are accessing the external Internet connection, so it is theoretically possible to "share" a small number of Internet addresses among all the users.

For example, a company may have a *Class C* address, which defines a subnet of 256 addresses (including the subnet address itself), to share to the Internet. Inside the access router, there may be 2000 to 3000 computers, but only 256 of them can access the Internet at any one time. This is the simplest form of NAT. The 256 "outside" addresses are considered a "pool" from which the router can draw to make a connection. The router does a one-to-one translation from an internal, private address to an external global Internet address. In this scheme, the connections are allowed to expire, so no user can hog an address. Although this is fairly simple in operation, the router must keep track of many details, so that it knows which internal address corresponds with which external address. Thus, this takes a more sophisticated (and expensive) router.

There is also a form of NAT that allows multiple computers to use the same address. This is a function performed by the simple routers that front-end a DSL or cable modem. These simple routers are often called NAT routers because that is their primary function. The NAT router must go to a "stub," which is a single subnet domain that has only the one connection to the global Internet. When a computer on the local network sends a request to a server on the Internet, the NAT router substitutes its own global Internet address as the source address. The response returns to that originating address, where the NAT router determines the identity of the local computer, substitutes its local address within the private subnet, and passes the packet to that computer. This is very similar to the function that a normal router performs, with the address translation added.

Choosing a Wired Hub/Switch

You will need an Ethernet hub or switch to allow multiple PCs and your wireless AP to connect to your router. Some SOHO routers are made specifically for sharing a DSL or cable modem and have a small hub built in.

An Ethernet hub is a simple device with four or more places to plug in Ethernet data cables. Ethernet has evolved considerably over the 2½ decades of its use. However, at this time, connections to your PCs and router are standard modular *10BaseT/100BaseT* cables. These cables are sometimes called Category 5e patch cords.[12]

Ethernet hubs connect different network devices together to form a simple network. Hubs are totally transparent to higher-level protocols, such as TCP/IP, so you don't have to worry about configuration. Basically, a hub repeats the data packets received from any port to all the other ports on the hub. It is like a big party line for data. You wouldn't worry much about security with a small hub on a wired network. The only other PCs that can listen in are within your small network. The wireless AP normally passes on only the packets that go to or from the WLAN devices. However, if you are attacked by a stealth Trojan that locates itself on one of the computers on your own network, keep in mind that the attacker will be able to see every packet that goes into the hub, no matter what the source.

An Ethernet switch is outwardly similar to a hub, but it switches each packet individually and sends it to the device with that particular packet's destination address. The switch learns the MAC addresses of your PCs, APs, and routers. You recall that an Ethernet packet has a header with the source and destination MAC addresses, in addition to the IP source and destination addresses. This is somewhat similar to the difference between your physical street address and your name. A letter could be sent to your office at 1234 Any Street. The carrier would deliver the letter to that physical address, where your mailroom would make sure the letter got to you. In the networking universe, all devices that connect to Ethernet have a physical or MAC address, but the software servers and clients have IP addresses. You could have a computer with more than one IP address, although an IP feature called *ports* more often handles that sort of thing.

An Ethernet switch has two primary advantages. First, it ensures a degree of privacy because the only device that sees an incoming data packet is the one for which the packet was intended. Second, two or more pairs of devices can talk through the switch simultaneously, which theoretically can multiply the effective bandwidth of your network. Remember that in a hub only one device can talk at a time, so you are limited to the maximum data rate of the hub ports. If you have a 10-Mbps hub,

[12]Lots of information on Ethernet, 10/100BaseT, and Category 5e is contained in *LAN Wiring* 2nd ed.

then that is the most that the hub can pass. If you have a 100 Mbps hub, then that is the throughput limit. However, an 8-port 100-Mbps switch can have four pairs of data transfers going on, which translates to an effective 400 Mbps throughput. There are two caveats here: devices often compete for the same port (such as a server or router) and you rarely get more than about 60 percent of the theoretical data throughput.

Do you need an Ethernet switch for your SOHO network? Unless you are doing lots of very intensive data transfers locally, such as massive video or graphic files, you probably do not need more than the 100 Mbps of bandwidth of the better hubs.

By the way, if you have a 10/100 switch, it may have the capability of operating its data ports at different speeds. For example, you could connect to some devices at 10 Mbps and others at 100 Mbps. To do this trick well, the switch must buffer (temporarily store) the packets and be able to restrain the data from the faster port. A PC that is connected to the switch at 100 Mbps can send data way too fast for a switch to pass to a 10-Mbps cable modem interface. Of course, another bottleneck would decrease the uplink speed of the cable modem.

Some SOHO routers have a built-in hub or switch. These are usually limited to four ports, so you will have to add another stand-alone hub/switch if you have more than four additional devices you want to connect to. Also, it is possible for the DSL/cable modem, the router, and the four-port hub/switch to be combined in the same unit. This is the most convenient situation of all, but it can be the most difficult to troubleshoot because you do not have the ability to isolate devices and substitute components. However, it is certainly the most compact and possibly the most cost-effective solution.

Backup Power

If you have a significant investment in your home network, you should consider adding backup power. None of the network components will operate without power, and power is always provided by the subscriber (except for the ordinary analog telephone). The cable or DSL modem, the NAT router, the hub, and the wireless AP must have power to function. Your laptop computer will keep working for at least a couple of hours on internal batteries, whether it is directly connected to the hub or connected wirelessly to the AP. In addition, all of your desktop computers will immediately turn off without power and reboot themselves when power is restored.

Electrical power may be interrupted by severe weather, rolling blackouts, or power failures. To minimize the effects of these events, you can provide a device called an uninterruptible power supply (UPS) to continue AC power to your critical network resources. UPS systems are available in a variety of capacities and configurations to match almost any requirement or pocketbook. As you would expect, the better UPSs are a little more expensive.

In a typical SOHO network, you should provide one UPS for the networking equipment and one for each AC powered computer. In this case, the networking equipment would include the DSL or cable modem, the router, the hub, and the AP. These devices typically do not take much power, so you might be able to power them from the same UPS used for one of your desktop computers, if it is nearby.

Project: Wireless Home Office Network

Our project in this chapter will show you how to create a simple, peer-to-peer network and supplement it with WLAN connectivity. This project is suitable for a SOHO installation. In the modern, computer-enabled home or office, there are often multiple computers that must share an Internet connection. Traditionally, we have been tethered to our desktop computers, just as they are tethered to the wall, for power and network connections. WLANs free us from these constraints and essentially make the Internet available throughout the entire home or office. The advent of highly portable computers, such as laptops and hand-helds, gives us the opportunity to do many of our computer tasks in the locations that are the most convenient or, in many instances, the most comfortable.

The wireless SOHO network will allow you to share Internet, printers, and file storage between several computers in your home or office, including those that are wirelessly connected.

What You Will Need

1 DSL or cable modem line
1 DSL or cable modem
1 SOHO router (possibly with the hub built in)
1 Ethernet 10/100-Mbps hub or switch (four ports, minimum)
5 Category 5e 10/100BaseT patch cables

Project: Wireless Home Office Network

1 Wireless AP
2 NIC, Ethernet 10/100-Mbps PCI adapter cards
2 W-NIC, 802.11b wireless PC adapter cards
1 PCI adapter for W-NIC

Project Diagram

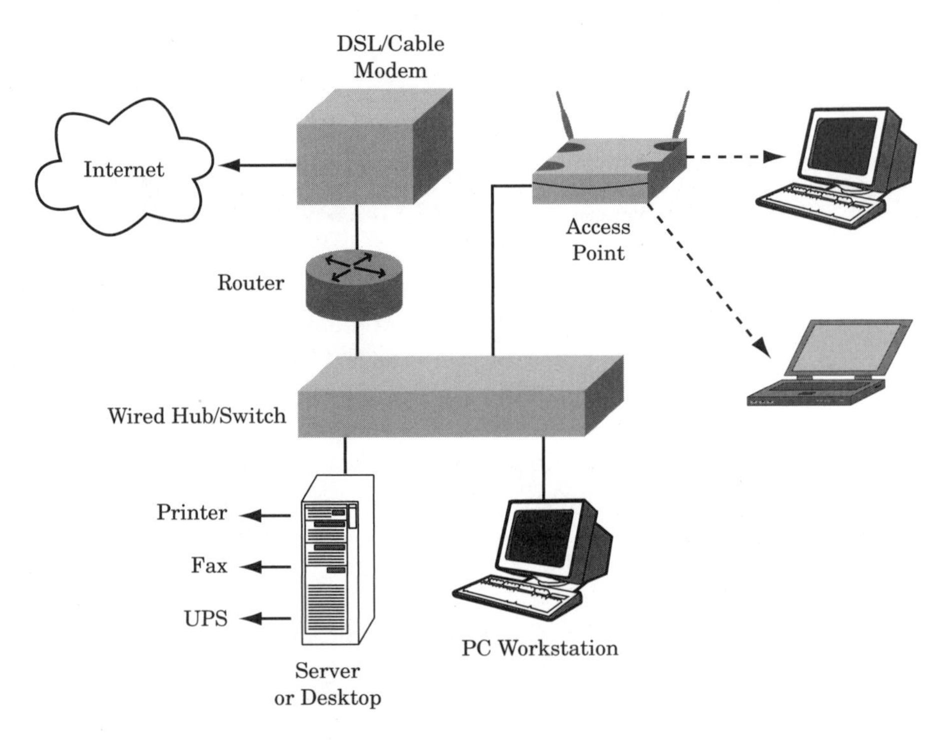

Project Description

In this project, we first construct a wired Ethernet network using 100BaseT (optionally 10BaseT) standards. The initial network consists of a 10/100BaseT dual-speed Ethernet hub, one primary desktop PC, one secondary desktop PC, a simple NAT router, and a high-speed access modem for Internet connectivity. The access modem may be a cable modem or a DSL modem, depending on your choice of access provider.

Project: Wireless Home Office Network

The primary desktop PC has a shared printer, fax printer, and file space. Windows peer-to-peer networking is used to allow the PCs to share these resources. All PCs have access to the Internet through the hub and router.

We then add the WLAN to the network. This part is a repeat of the project in Chapter 3, Simple Wireless Home Network, so this description is abbreviated. You can refer to the previous project for additional tips on constructing the wireless portion of your home office network.

We assume that you have at least two desktop computers for the wired portion of the network and one laptop and one desktop to be connected via a WLAN. You may have more or fewer computers for each function; if so, you should adjust the quantities of NICs and W-NICs listed above. Let's get started.

Step by Step

Step 1: Install the 10/100 Mbps Ethernet Cards

For this step, you should prepare the desktop PCs that are to be wired directly into the hub. Specific instructions vary with the brand of PC and the type of NIC you have. You can find specific details in the PC manual (or help files) and in the manual provided with the NIC. In general, you will have to turn off the PC, remove the power cord (some PCs do not shut down completely), remove the case side (on a "tower" case, the left panel as viewed from the front), remove the metal knock-out panel for the adapter slot you wish to use (be sure to test-fit the PCI card to see which panel to remove), and place the card. Be sure to review the cautions in Chapter 3 about handling this card to protect it from static electricity and properly seating the card in the PCI connector. Then you should reassemble the case and reconnect the power.

When you power on your computer, you will need to install the drivers for the new Ethernet NIC. The drivers will be supplied on 3.5-inch diskette or on CD-ROM. Most of the Windows 9x, ME, and XP OSs will have plug and play enabled, and will recognize that new hardware has been added. As Windows is booting, when the New Hardware Found alert pops up, choose "Have Disk" or "Driver from disk provided by hardware manufacturer" (Win 95). You will be prompted for the location of the drivers, which will be the A: drive for diskettes or the D: drive (sometimes E:) for CDs. As soon as Windows finds the drivers, it will display one or more product drivers that are on the disk. Your adapter should be on the list. Highlight it and click Next or OK to proceed.

The NT-based (NT, 2000, etc.) and Linux OSs will have to be manually configured. You should go to Control Panel, choose Add Hardware, and proceed in much the same manner as with Windows 9x systems.

After the driver is installed, you will be able to configure your network settings. *Be sure to have your Windows 9x (or the appropriate) Setup Disk handy before you begin the next steps because you will need to insert the disk to complete any changes to the network settings. (Some factory-installed systems will have the Windows Setup files already on the hard disk.)* For a new network installation, go to Control Panel and choose Network (the instructions for XP are slightly different, as covered in Chapter 3).

On the Identification tab, enter a computer name (for your own reference) and matching description. For example, you might want to use the brand of computer, your initials, or some easy identifier. This is the name by which your computer will be known when you remotely connect to it for access to its printers or files. Next, enter a unique workgroup name, preferably something with letters and numbers that is fairly long (you are permitted 15 characters), to enhance workgroup security. You never have to type this again (except to assign the other computers in your network to the same workgroup), so you can be just as obscure as you want to.

Now click the Configuration tab and then the File and Printer Sharing button. If this computer is used for shared printers and file storage, check "I want to give others access to my files" and "I want to allow others to print to my printer." Click OK to close this dialog. On the Configuration tab, look at the list of installed network components. Your newly added Ethernet NIC should be listed as an adapter (with a NIC card option). Unless you previously installed networking, you will need to add the TCP/IP protocol. In turn, select Add, Protocol, Microsoft (under Manufacturers), and TCP/IP. When the protocol is added, highlight it in the list of installed components and click Properties.

The Properties settings will be the same for most standard DSL/cable network installations and are independent of the Windows workgroup settings, in most cases. In the TCP/IP Properties dialog, on the IP Address tab, select "Obtain an IP address automatically." Make the following settings on the respective tabs: WINS Configuration tab, "Disable WINS Resolution" (unless you have to connect to a remote company network); Gateway tab, normally blank; DNS Configuration, normally "Disable DNS"; NetBIOS Tab, uncheck "I want to enable NetBIOS over the TCP/IP protocol," if you can, for security reasons; Advanced tab, uncheck "Set this protocol to be the default protocol," if you can, to attempt to force NetBIOS to support Windows networking.

On the Bindings tab, you may be able to uncheck all the bindings because the Client for Microsoft Networks and File and Printer Sharing for Microsoft Networks may be handled over NetBIOS as a default. (When you leave this tab, you will be alerted that you have not chosen a client to bind to this protocol, and asked if you want to choose one now. Choose No.) This setting is a benefit for network security. However, some OSs, particularly ME, seem to balk at this setting, presumably because they want to allow encapsulation through TCP/IP. In these cases, you may be forced to allow the bindings, and perhaps to set TCP/IP as the default on the Advanced tab.

Click OK to close the TCP/IP Properties dialog and OK to close the Network applet. The system will scan to build a driver information database and may prompt you to insert the Windows 9x or ME disk. Insert the CD and help the system find the right drive letter if it is confused. Next you will be informed that the system needs to be restarted to enable the changes. Choose Yes, and your computer will reboot with the new network features.

Note that you will not receive a valid IP address, unless you are connected through your hub to your DSL/cable modem router. See steps 2 and 3.

Step 2: Connect the 10/100 Mbps Ethernet Hub and Create the Peer-to-Peer Network

The Ethernet hub allows all the computers, the router, and the AP to talk together. You need to connect standard Ethernet cables from the hub to all these devices. It is a good idea to locate the hub in a central location, with the DSL or cable modem and AP nearby. If you have a typical home, with no special data cabling, you probably will want to install the DSL/cable modem in the same area as your main PC (which we call the primary PC in this project) to keep most of your cables relatively short.

Some homes have a centralized telephone and data wiring system. In this case, the best place for the DSL/cable modem, the router, and the hub is at the wiring distribution panel—typically in a utility closet or in the garage. Residential cabling is becoming more prevalent, and you may wish to add this feature to an existing home. With centralized cabling, you will need additional Category 5e patch cords because you need to connect to the computers throughout the house and interconnect in the central location to the appropriate data cable for each room. If you have centralized cabling, you probably will have centralized CATV distribution as well, so both telephone and cable TV lines will be terminated in the same location as your modem, hub, and router. By the way, this is a perfect location to place a backup UPS to maintain telephone and data service in a power outage.

Connect the patch cords between the Ethernet hub and the router, the wireless AP, and each wired computer. You should have a green light by each connected port on the hub to indicate a connection. You should connect the router's cable to port 1 on the hub. In some cases, you will need to flip a small switch to *cross-over* that port, or there may be an alternate port 1 (usually indicated by an "X," "Crossover," or "Concatenate;" some hubs have this feature on port 5). Some routers are designed so that a single PC can plug in and use this cross-over capability to reverse the cable. This is so that the transmit and receive pairs from the router go to the correct pins on the hub. The router usually will have a LAN light to indicate a proper connection from the hub. On your PCs, you should also see a green light on the Ethernet PCI card panel, although it can be a chore to get around behind your PC to look. You could proceed and see if it works before trying this contortion.

If you have a router with a built-in hub, you will need fewer cables and not have to worry about the cross-over problem. Also, some hubs (including those built into routers) are Ethernet switches that provide extra security and, potentially, extra bandwidth.

Once you connect two or more computers through the Ethernet hub, you can try out your peer-to-peer network functions. We did the basic network configuration and setup in step 1. If Windows networking is working successfully over NetBIOS and not encapsulated over TCP/IP, you should be able to double-click Network Neighborhood (or My Network Neighborhood) and see icons displaying the other computers in your workgroup that are connected to the hub. However, if your OS wants to force NetBIOS over TCP/IP, you will need valid IP addresses on each computer. Remember, these IP addresses are assigned by your SOHO router. When a PC is booted, it will request an address (using DHCP) from the router, so the router should be turned on and connected to the same hub as your PC.

If you have trouble getting Windows Networking to function without TCP/IP, skip to step 3 to get your router set up and then return to this step.

Now you can set up the shared resources on your peer-to-peer network. Keep in mind that this function is independent of normal Internet protocol and supports only extended Windows functions, such as file sharing and printer sharing. The first item to share, in our project, is the printer connected to your primary PC. First, we must "share" that printer. From the Printers folder (open Start/Settings/Printers or double-click My Computer and Printers), right-click on the printer you want to share, and choose Sharing. (Obviously, that printer will have to be installed on your primary PC and should be turned on.) On the Sharing tab of your printer's Properties dialog, select Shared As and enter a Share Name. This name should be descriptive (such as HP-Laser or Laser-on-1) so that you can easily recognize it from the printer list that appears in your applications. Add more shared printers and the fax machine. Notice that each of the printers will have a hand added to the printer icon to indicate that they are shared.

From the second PC, choose Add Printer (from Start/Settings/Printers or you can double-click My Computer and Printers) to activate the Add Printer wizard. Choose Network Printer and then choose Browse to look for the name of your primary computer. Double-click on the name or click the "+" next to the name to display the printers that are shared. (This is where good share names are really important.) Choose one and click OK. Follow the instructions on the following screens and print a test page. Before printing the test page, the second PC will download the proper printer drivers from the first PC and set up your shared printer. If you have no other printer connected, you should select this printer as your default printer. If you did not do this from the printer wizard, you can right-click the printer in your Printers folder and choose Set as Default. The printer icon will now include a check mark to indicate that it is the default printer.

To share a file space, you must first have file sharing enabled in the sharing (target) PC's network setup. Then you need to open Windows

Explorer and right-click on the folder you want to share. If it is the entire C drive, right-click on C:. Choose Sharing, set the Shared As name, and select Full. Enter a password that you can remember. You will need to enter this password again the first time you try to access this shared file space from another computer.

Step 3: Add Internet Access with DSL/Cable Modem and Router

Step 3 is fairly simple, but crucial, because it gives all of your locally connected PCs and devices an IP address and it connects you to the Internet. First, you should connect the DSL or cable modem to the DSL or cable line. Be sure you have a modem with an Ethernet interface. Some DSL/cable modems are intended for connection only to a single PC through the USB port and will not work in a SOHO network. Many of the newer modems have both USB and Ethernet ports, so you can do either. The modems connect to their respective lines somewhat differently.

A DSL modem connects to the DSL jack over an ordinary-looking telephone cable. Be sure to use the DSL cable supplied because some DSL modems connect on the outer two pins of the jack (an RJ-11–style 6-pin jack). You may have received the modem with a self-installation kit. The installation kit will contain the modem, an Ethernet cord, sometimes a free NIC card, and several telephone filters. The filters allegedly keep the DSL signal out of your telephones. It is rare, but possible, for the DSL signal to interfere with the operation of a telephone. Remember, both services operate over your home (or business) phone line. More likely is that many telephones on your line will decrease the DSL signal, if you don't use the filters. If you have a centralized phone cabling system, you may want to request a line filter that is designed to pick off the DSL signal at the point of entry (commonly called the *demarc*, or point of demarcation). This is better than having lots of annoying little filter boxes crammed into one place. You may need to run a "clean" line from the service entrance (probably a gray plastic box near where the phone line enters your house or building). If you have an apartment or condo, you probably have a location where the main phone line comes into your unit, perhaps in one of the telephone jack boxes.

The cable modem must connect to a "clean" cable TV line to make its connection. Cable modem signals are sent and received on the same cable as your TV signal but may be more sensitive to levels and reflections. Splitters, amplifiers, connectors, and older coax cable may affect the power level of the signal. Reflections may occur from bad connections, cheap splitters, or even unterminated splitters. The cable company may want to run a new cable from the point of entry at your house or office back to your cable modem. If you have a centralized system, be sure to have this cable run to that location. In recent years, cable companies have sent out kits that include the cable modem and setup software to keep from having the labor cost of a "truck roll." They are much more likely to try to connect you with a USB modem, so be sure you get one with an Ethernet interface. The proper way to answer the ordering agent's questions is "I have a Windows 95 computer without USB, so I will need an Ethernet modem." If they ask if you already have an Ethernet interface, say "yes," although you can accept a free card if they offer it. If you say you have 98 or XP, they may send you a USB modem that will be useless for your network project. Fortunately, many newer cable modems have dual USB and Ethernet interfaces, so you won't get stuck with a USB-only modem.

The project diagram shows that the ISP line (cable or DSL) connects to the modem, the router, the hub, and the PCs and other devices. The ISP automatically assigns the IP address (through the modem) to the router using DHCP. (In rare cases, the ISP will use a static address, and you will have to use the configuration screens in the router to put in that permanent address.) The router in turn assigns IP addresses to each PC and the AP, also using DHCP. This could create a problem if one of the devices in this chain wakes up without being able to get its IP address, or if the "lease" on the address expires and is not renewed. This often happens when one device in the chain loses power or is reset.

To make sure all the devices get a valid IP address, you will need to power up or reset them in a certain order. The basic guideline is to boot up "outside-in." Boot up the devices in the order that they contact the outside world. Thus, you need to boot up the DSL/cable modem first, then the SOHO router, then the hub if it is a switch (simple hubs don't need to be reset), and then the wired PCs and the AP. If there is no reset button (which is common), you can switch off power to each device in turn. You may need to unplug the DC power cord or the AC adapter. Keep each device off for a minimum of 10 seconds (counted slowly). A few devices may need to be powered off for a minute, but this is very rare and probably not necessary.

Now, you can make sure your PCs have valid IP addresses. For the PCs (and the AP, if it has an IP address), this address will be within the local subnet, which is controlled by the router. A typical address could look like 192.168.1.101. If you look at your PC's IP configuration (run WINIPCFG, or look at My Network from Windows, or run IPCONFIG from DOS), you will see a similar address (maybe the .102 or .103) and the gateway address of the router (such as 192.168.1.254). If the router has picked up a valid DHCP address through the cable modem, you will also see DNS entries, which your browser needs to use to find stuff on the Web. If you have any problems here, nothing else will work, so go back and troubleshoot your IP setup and connections. As a last resort, you might suspect a hardware problem with the NIC, the hub, the router, or the modem. However, most likely, it is a bad configuration on the PC or perhaps the router.

At this point you can connect to your router to check its setup, and by all means, change the default passwords. Default passwords are the simplest road to a security breach, and you should change them to a unique, but easily remembered, password immediately. Be sure to keep track of the new password. It is a good practice to keep your passwords together in a safe place, if you have trouble remembering them. A lost password can cause a world of problems, as you can imagine.

You should now be able to connect to the Internet from any of the wired PCs. This is a good check to see if you have everything connected properly. If you have problems, some good troubleshooting advice is: divide and conquer. Try to isolate what works and what doesn't. For example, if Windows networking is OK, then your NIC card, NIC drivers, hub, and the connecting cables are also fine. If you can do the setup/configuration on your router using your browser, then your TCP/IP configuration is likely fine, and the router is properly connected to the hub and OK as far as the local Ethernet side is concerned. If the router has a valid global IP address from the ISP, then the WAN side of the router is OK, as is the modem. If you still can't browse, you may have a problem with the DNS settings. In general, you want the ISP, and not the router, to provide DNS, so check that setting. Also make sure that the PCs' configurations are set to get DNS from DHCP, which usually means that local DNS is disabled. If all else fails, its time to get your ISP or possibly your router manufacturer on the phone to assist you.

Step 4: Install the Wireless AP

Now, let's go wireless! Connect the AP to the hub, if you have not already done so, and turn the power on. You should get green lights on the hub and on the AP to indicate a good connection on both ends. Your AP model may allow configuration from the Ethernet side, in which case you can install the AP management software on one of the PCs connected to the hub. In rare cases, more advanced hubs may allow setup over a serial cable directly connected to a PC. However, it is more likely that your SOHO AP will come with the proper defaults and not really need any configuration at first. Defaults are defects and should be replaced with real passwords and group settings as soon as possible. In the case of 802.11 networks, one of the important settings is the network name. By changing this, you make it more difficult for a snoop to break in.

You should determine the best location for the AP based on the desired coverage range of your wireless network and anticipating the effects of walls, windows, floors, and metal structures, such as air ducts and appliances. You may be able to locate the AP in another room from your primary computer if you have centralized cabling. This is a good time to consider the mounting height for the AP, or for an external antenna, if your AP has that capability. The best way to test the AP is to complete step 5 and connect to it from a wireless PC.

Step 5: Install W-NICs in a laptop PC and an adapter in a desktop PC and connect to the network

Install the W-NICs in your laptop and desktop PCs by following the instructions in Chapter 3. We have specified one wireless desktop PC in our project so that you could experiment with installing a PCI adapter card for your W-NIC. You will need to physically install the cards and then install the driver software. In at least one of the wireless PCs, you should install the AP management software.

After installation of these components and the required reboot, bring up the management software to verify that you can detect the AP. You can also verify that you have adequate signal strength. Remember that the 802.11 network degrades to progressively slower connection speeds to give you the greatest chance to connect. If you do not have enough signal, you will need to move the PC or the AP or adjust the antennas, if possible.

Once you have a solid signal between your wireless laptop and desktop and the AP (remember, this is a two-way connection), you can check the network setups on each PC. Use the same procedures in steps 1 and 2 to add the TCP/IP protocol (if not already installed), configure TCP/IP

(DHCP, DNS, and NetBIOS), and set your workgroup ID. You do not need to allow printer and file sharing on these PCs unless the other wired or wireless PCs need be able to use their files or printers (a possibility for the wireless desktop PC).

Try to "see" the other computers in your Network Neighborhood (or My Network Neighborhood). Remember, the workgroup names must be identical in all PCs in the shared network. The peer-to-peer networking can operate independently of TCP/IP, so you might be able to get to the Internet but not your peer network, if everything is not set up correctly. You should be able to see the primary PC and its shared files. You will have to type in a password the first time you make this connection if you password protected the shared resource (which you should have). Peer networking works best with the same OS versions on each PC. As we have said, ME has peer networking issues, and you may have to resort to encapsulating NetBIOS over TCP/IP to make it operate with non-ME PCs.

The final step is to connect to the Internet from your wireless PCs, if you have not already done so. You should be able to bring up your Web browser and see your default home page. (If your PC has been configured to *always* dial up for the Internet connection, you will need to modify this setting. Go to Start/Settings/Control Panel/Internet Options and choose "Never dial a connection" or "Dial whenever an Internet connection is not present.") To see how well the connection works, you can bring up a known fast site, such as www.yahoo.com or www.usatoday.com. You can also check your wireless speed by looking at the signal strength shown in your WLAN monitoring application. Depending on the manufacturer, some monitors also show the current WLAN link speed. WLANs "fall back" to lower operating speeds to compensate for lower signal strength and to reduce errors.

If you are not in an ideal location relative to the AP, you may be connecting at a lower speed than the 11 Mbps of 802.11b or the 22 Mbps of 802.11e. Fallback speeds are 11, 5.5, 2, and 1 Mbps, which is potentially a 20:1 difference. Keep in mind that the 1-Mbps speed is still comparable to the nominal 800 kbps (0.8 Mbps) download speeds of most DSL and cable modem connections, so access to the Internet may be very similar to those of the wired PCs. However, if you have a great high-speed connection over your DSL/cable line, you may see times at which your download speed will be several times the expected minimum rate. In such cases, the WLAN connection may be slower if at a fallback speed.

Applications that connect over the network to a server or a shared file space also will be sensitive to the connection speed. If you have 100-Mbps–capable NICs and hub, the very top speed of your WLAN (11 or 22 Mbps) will seem a little slower than the wired connection. However, most of us use applications that reside on the local PC and, at most, use the LAN for file storage. If reading and writing large files seems to be slowing you down, try using the central PC only for file backup and keep the large files local. Or you may want to switch to a true client/server application, which will probably require a true network OS, such as NT or Linux, rather than depending on your peer network.

Project Conclusion

At the conclusion of this project, you should have a dual wired/wireless network for your home or small office. You will be able to access the Internet from both types of PC connections, and you will be able to share printers, fax printers, scanners, and file storage from any attached PC.

Virtual Office from Your Easy Chair

Chapter Highlights

- Using a Wireless Laptop
- Sharing Files and Printers
- Wireless Phones and 802.11
- Internet Phone (Voice over IP)
- Project: Wireless Virtual Office

Once you have completed the projects in Chapters 3 or 4, you will be able to connect to your SOHO network with your WLAN. A WLAN is a wireless extension of a conventional LAN. Even if you do not have much of a wired network, such as the one we built in Chapter 4, you technically have a wired LAN, even if the extent of that LAN is just to connect your wireless AP to your DSL or cable modem.

Most of us will have one or more computers that will be connected to our SOHO LAN, and it is likely that we will have network resources, such as printers and disks, on one or more of those computers that can be shared with our wireless computer(s). This chapter is about using those resources to create a virtual office anywhere in your home or office—even your easy chair!

In this chapter, we expand our SOHO network project to employ shared resources, such as a printer or a fax machine, and to use voice over IP to place telephone calls. This Wireless Virtual Office project will bring home the power of wireless networking.

Using a Wireless Laptop

In many ways, using a wireless computer is no different from using a computer connected to a wired LAN. A wireless PC is interconnected to the network through a wireless link with the use of a standard wireless protocol, such as 802.11. There are, however, some special considerations involved, and this section covers those details.

One of the first things you will notice is that there is an extra item in your list of things that could go wrong: the wireless link. Therefore, it is important to thoroughly understand what features the WLAN adds, and how to troubleshoot any problems. Most of this involves stepping logically through a mental list of how you connect to the resources on your wired LAN or Internet connection. Once identified, problems are usually very easy to solve. Don't jump to conclusions just because you can't get your wireless laptop to connect to the Internet. First, determine whether you can connect to any of the other resources on your internal wired network. If so, then ask: Can any one of your wired PCs connect to the Internet? If it cannot, then the problem has nothing to do with your WLAN, and you should work on getting your Internet connection back.

Remember that your wireless laptop gets its IP address "automatically" from the router. Normally, it gets this address when it boots up. If you use the "sleep" feature on many laptops, when you turn it on, it returns to the same state it was in before you put it to sleep. This is not the same as a boot up or a restart! When your PC awakens from sleep, it maintains the address it had before. If that address if no longer valid, because you used the laptop at work or the DHCP address's lease has expired, then you will not be able to connect. Either releasing and renewing your IP address lease, or a quick reboot is required to renew your address and let you communicate to your shared resources, such as the Internet connection, through TCP/IP. Check out the Tech Tip on Renewing Your IP Address for more information.

Tech Tip: Renewing Your IP Address

IP addresses that are issued through DHCP are issued for a limited time. The idea is that IP addresses are a limited resource, and you should get only one for as long as you continue to use it. In the real world of global Internet addresses, this view is quite accurate. Your ISP may have fewer IP addresses than clients, and this feature allows the addresses to be reused if someone is not active on the Internet connection or if they drop their service. Also, this mechanism allows an orderly reuse of addresses, so that there are no holes in the address space.

In your SOHO network, however, your PC is very likely getting its IP address from your router, from the pool of private addresses in your local subnet. For example, you may be using the private subnet 192.168.1.0/24 with 255 addresses all to yourself. There is absolutely no reason to ever have to give one of these addresses up, other than that there might be some confusion if two of your computers thought they should use the same address. The renewal process takes care of that, as does a reboot. In each case a "fresh" unassigned address (which may be a reissue of the same address, now released) is requested from the DHCP server portion of your DSL/cable router. That address is issued with a finite shelf life, called the "lease."

If you look at WinIPcfg (pronounced Win IP config), calling up Start/Run and typing in the name "winipcfg" in a Win 9x/ME system, you will see your current IP address. You have to choose your Ethernet NIC from the pull-down list. This dialog shows your Adapter Address (a unique 12-digit hexadecimal MAC address), your current IP Address, the Subnet Mask (normally 255.255.255.0), and your Default Gateway (probably 192.168.1.1 or .254, the address of your router). You can see the Release and Renew buttons on this dialog. More on that in a moment. If you click the More Info button, you will see two parameters, the Lease Obtained and Lease Expires values. These values show when you initially requested the IP address, and when your computer will have to request that the IP be renewed. Your router supplies these values.

Your SOHO router has similar IP lease values for the global IP it obtained from your ISP. The ISP's router may issue DHCP addresses for periods as brief as 1 hour. In some cases, a very quick expiration of the IP address is responsible for significant performance issues, but in most cases the time will be long enough not to be troublesome. Most ISP's set their DHCP lease time to exactly 1 hour, 4 hours, or 24 hours. That means that you could get a totally new IP address at the end of that time. It is more likely that this would happen if your computer was off when lease renewal time came. However, your SOHO router is probably connected all the time, so it will renew at about the same time each day.

For your private network, your router may use the 24-hour default or it may go 5 or 6 days. You may be able to adjust this setting, but there is no strong reason to do so. After all, you have the other 253 addresses to play with at any time, so what is one address?

If you find that you frequently need to manually renew your IP address, you can put a shortcut on your desktop to WinIPcfg. First, display your files in Windows Explorer. Go to the Windows directory and find the file called "winipcfg.exe" (the .exe may not show, if you hide known extensions in your folder views). Now, with the right mouse button, click and drag the icon for WinIPcfg onto your desktop. Choose Create Shortcut from the pop-up menu, and the icon will appear on your desktop. Rename it "IP Config," if you like, so the whole name will show.

Here's a clever trick if you have Win 98 or higher: you can put WinIPcfg in your Quick Launch bar, next to the start button. Simply right click and drag the new IP Config icon from your desktop to the part of the taskbar ribbon that is next to your start button. You will see a vertical cursor bar appear, and you can position the icon with that cursor wherever you want on the Quick Launch Bar. Release the mouse button, and you will see a miniature version of the icon on the task bar. If you have more icons than space, a double arrow at the right of the Quick Launch bar allows you to see the other icons. Now you can single-click the IP Config icon anytime you want to see your current IP or renew your IP. This is a great time saver, because you don't have to wait for your PC to reboot.

If you have XP or one of the NT systems, you will have to go through a more manual process to renew your IP. It is too bad that the developers left this handy applet out of those systems.

Connecting to Resources

You will connect to several types of shared resources from your wireless laptop. Obviously, the router connection to the Internet will be shared. In addition, you will have the ability to set up shared file space and shared printers. As a matter of fact, virtually anything that is connected to any of your wired (or wireless) PCs can be shared, as you can see from Figure 5.1.

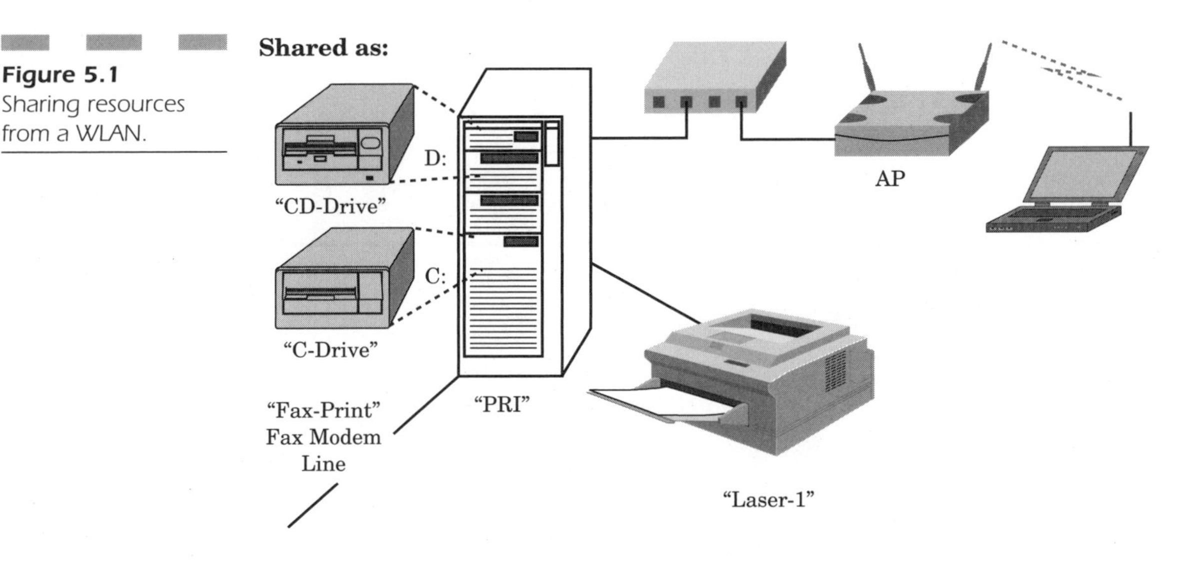

Figure 5.1 Sharing resources from a WLAN.

For example, you might have a handy CD-RW drive, or so-called CD writer, on your primary PC. A CD-RW normally has a drive letter associated with it, and you can share that drive in the same way you share your C drive or any of your backup folders. Let's start with file sharing and then cover sharing those other resources.

To share a disk file folder (or directory) on one of your PCs (the target PC), you must complete the following steps:

1. Enable file sharing in your Network configuration on the target PC.
2. Enable sharing for that particular folder or drive.
3. Access the shared folder from another PC through Network Neighborhood or through mapping a network drive.

To enable file sharing on the target PC, click the Start button, choose Settings, and double-click the Network icon. (We will abbreviate this procedure as Start/Settings/Network or similar sequencing for future call-ups.) Alternatively, you can right-click on the Network Neighborhood icon on the desktop and choose Properties. For the next portion, you must have your Windows Setup disk available, either the CD-ROM or available on the the hard drive. The Network configuration applet window appears (Figure 5.2). Click File and Print Sharing, and check "I want to be able to give others access to my files" (Figure 5.3). Click OK twice, and, unless file and printer sharing are already installed on your computer, you will be asked for your Windows disk and then to restart your computer, which you should do.

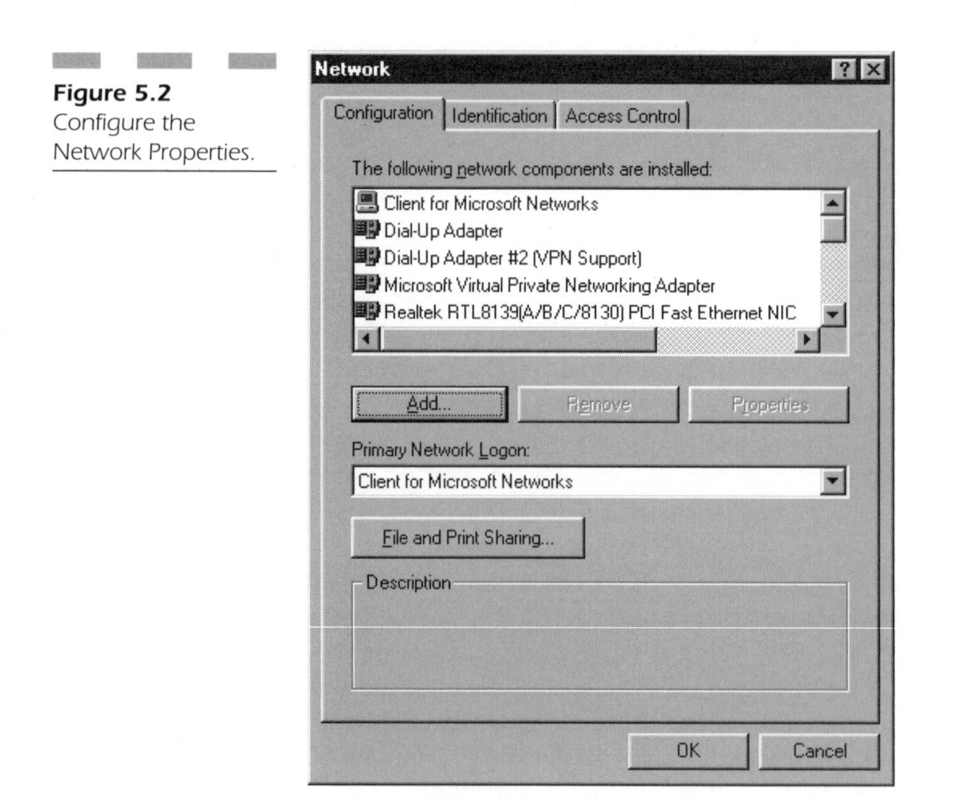

Figure 5.2 Configure the Network Properties.

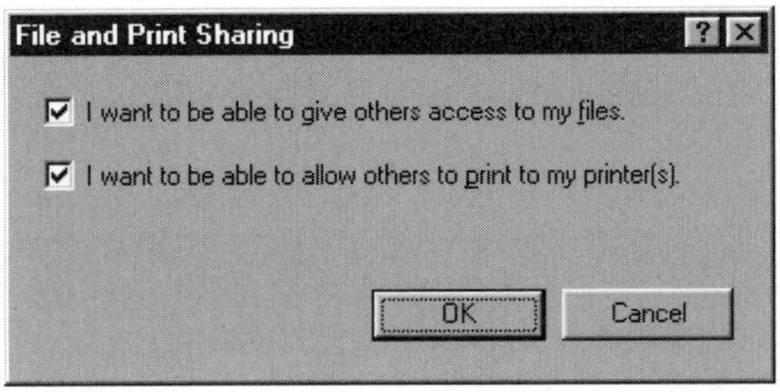

Figure 5.3
Select File and Printer Sharing.

File and Print Sharing

I want to be able to give others access to my files.

I want to be able to allow others to print to my printer(s).

OK Cancel

So far, only the capability to do file sharing over Windows Networking had been enabled. Now we need to enable sharing for each resource we want to share. For the file folders, you should open Windows Explorer to show your list of drives and folders. You will notice that Windows Explorer is divided into right and left frames. The left frame, titled Folders, shows all of the folders associated with your computer, whether they are drives, file folders, your local network, or printers. The right frame shows each of the files or devices within the selected folder. You select a folder by clicking once on that folder's name or icon. (Careful—if you leave the mouse button down or click twice slowly on the folder name, it will change to a Rename text box, and anything you type will replace the existing folder name. Press Escape to back out of this.)

The folders that will be of interest to us in setting up our shared resources are the C: drive, the D: drive (or the correct drive for the CD-RW), the Printers folder, and the Network Neighborhood. You can also access these drives and folders through the My Computer icon on your desktop, if you want to, by double-clicking on each icon in turn until the one you want to share is displayed.

To share the entire C: drive, right-click on the C: drive icon and select "Sharing...." The ellipsis is Windows' way of saying "more to come." This displays the C: Properties dialog box, as shown in Figure 5.4. Select Shared and enter an appropriate name to identify your shared drive to the network, in this case "C-Pri"[1] for the C: drive on the Primary PC. Select Full for the Access Type and enter your Full Access Password (maximum of eight characters). You will be asked to confirm the password. Make a note of this password, so it will be easy to remember when

[1]Some punctuation characters may not be allowed, as with 8×3 filenames. If you use a longer name, be sure to type this name with no spaces because it will confuse some software. You can usually use a dash or underline character as a word separator. Also, the name and password are case sensitive, so pay attention to capital letters.

you access this resource from another computer. It is not really critical because you can always reenter it. However, it is annoying because all of the computers that formerly had access to the shared resource will have to reenter the new password.

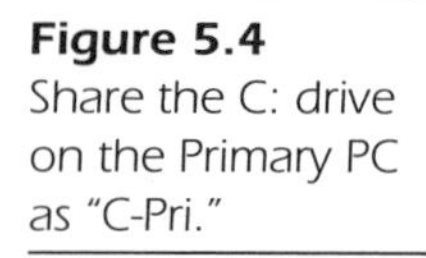

Figure 5.4
Share the C: drive on the Primary PC as "C-Pri."

You may also change the access levels between Read-Only and Full by specifying "Depends on Password" and entering read-only and full access passwords. This might be useful if you wanted someone to have access to a particular document, such as a calendar or telephone list, but not be able to change it.

The entire C: drive does not need to be shared. As a matter of fact, there are some security and privacy issues you might want to consider. Basically, anyone who has full access to your C: drive could read, write, or even erase it without your permission. You might think that, if you are the only user, this would not be a problem, but any process on the other computer has privileges just like a real, live user. If you have a bad virus or a stealth Trojan on another computer, it would have complete access to your primary computer and could do anything that was possible on its local drives, including erase them!

A more conservative approach is to have a shared area of your C: drive, perhaps even a subdirectory within the folder My Documents, where you explicitly place documents that you want to share. For example, you could call this folder My Documents\Shared. You would move or copy a document to this folder to make it available on the other computers. Alternatively, you could move or copy a file from your wireless laptop to this folder on your primary PC so that you could archive it, print it, or have it available to the other computers in your network. In an office, this is a fine way to share documents, without giving everyone the keys to the store.

To share a folder on your Primary PC, go to the file manager, Windows Explorer,[2] and right-click on the folder you want to share. Select "Sharing..." and complete the steps above, just as you did when sharing your entire C: drive. Sharing a folder implicitly shares all of the contents within that folder, including subdirectories (or subfolders) and shortcuts.

Now that you understand how to share the C: drive, you can apply this same process to any of your other resources, except for printers, which we cover in the next section. For example, to give access to a Zip drive or CD-RW, right-click on that resource, assign it a share name, access type, and password, and you are ready to go. This works great for moving data to a permanent backup disk or for loading software from your central computer. You could also share a FlashCard adapter for your photos or share a scanner. Anything that is connected to a computer on your network can be shared in this manner. The only problems you may have are with devices that are intolerant of delays. The process of converting data packets to and from WLANs, in some cases, can produce minor delays in data transfers. Some applications, such as copying an audio CD (appropriately called "ripping off"), transfer massive amounts of data very quickly. Such programs will not work well over a network, whether it is a WLAN or a conventional one. Copying to a data CD-RW, however, usually works fine.

---

[2]To save time, you can place Windows Explorer on your Quick Launch Bar (in Win 98 and higher). If the icon is on your desktop, right-click the icon and drag it to the Quick Launch Bar portion of the Task Bar (at the right of the Start button). If the Windows Explorer icon is not visible, open Start/Programs, right-click on the Windows Explorer item, and drag it to the Quick Launch Bar. When you release, you will be prompted for an action. In both cases, choose Copy (since this is already a shortcut). If you want to cancel in the middle, drag the cursor around until you see the "Not" symbol, ⦸, and release the mouse button.

Accessing Shared Files

Once you have set up shared file space on your primary PC, you need to know how to access the file folders from another PC on your LAN. This procedure is identical for wired and wireless PCs. As with everything in networking, there is a difficult way and a cumbersome way. Well, actually, setting up remote file access is pretty simple to use, once configured. What we can do is to set up the remote shared folder as if it were a local drive (resource).

To set up access to a shared disk or file folder on another computer, you must complete the following steps:

1. Display the shared resource in Network Neighborhood
2. Map the resource as a local drive
3. Correctly enter the password for the shared resource

For the shared folder in this example, let's use the Primary PC's C: drive that we shared in the previous section. Its resource name is "C-drive" with no spaces. The path description is "Pri\\C-drive," indicating that it is a folder on the network device "Pri." On the laptop PC (or other secondary computer) open Windows Explorer (not to be confused with Internet Explorer, which is a browser). The left frame shows all folders (another name for "folder" is "directory"). This frame is easier to view if the directories and subdirectories are hidden. Click on the "–" sign to the left of the C: drive label to condense and hide all the directories in C:. Scroll down in the left window until you see Network Neighborhood (Figure 5.5).

Double-click the Network Neighborhood item (or click the "+" sign to the left of the label) and on the workgroup name to display the other computers in your workgroup. You should see the Primary PC in the folder tree under the workgroup name. If you do not, or if you get an error, you do not have Windows Networking configured properly. Go back over the procedures in Chapter 4 to properly configure Windows Networking.

If you have properly set up sharing for the C: drive on your Primary PC, you will see an additional item underneath the computer icon for Primary PC, with the share name Pri. (Remember, use no spaces if you use a more descriptive name, such as PrimaryPC.) You could also have used something like "PrimaryC" to distinguish that this is the entire C: drive on Primary PC, but this is fine.

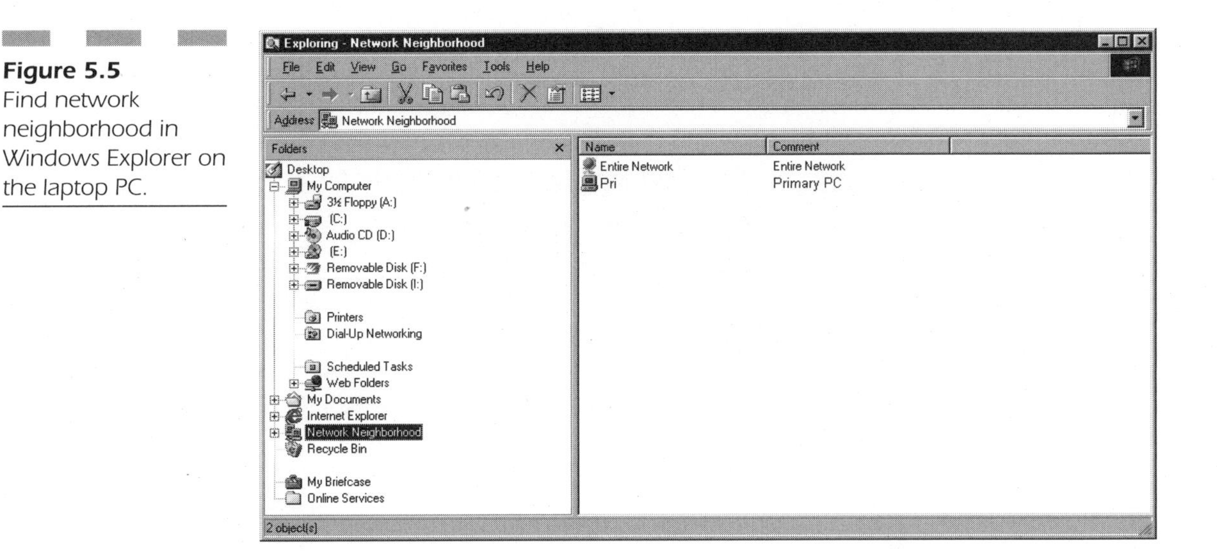

Figure 5.5 Find network neighborhood in Windows Explorer on the laptop PC.

Now we will do the difficult and cumbersome part. If you click on the shared resource, PrimaryPC, in the left frame of Windows Explorer, you will see all the directories in the C: drive on Primary PC displayed in the right-hand frame. You can click down through the levels until you see the folder you want and then use drag-and-drop to move the files to or from your PC. Remember, your own folders are still available to you in the left-hand frame. That's a little cumbersome, and really confusing, so it is very easy to make mistakes that take hours to figure out.

Let's try the difficult process and see if it really makes things easy. Let's make an area of our Primary PC appear as a local resource, namely a new drive. Suppose we set up a new folder on the Primary PC called "My Shared Files." That is convenient because Windows Explorer alphabetizes its folder list, so this new folder will appear right below My Documents. To do this properly, we could go back to Primary PC, remove sharing from the entire C: drive, and enable sharing for My Shared Files, giving it the share name "PriShare" and a full-access password of "pri123." Note that passwords and the share name are case sensitive. This won't be a problem for the share name because we are going to point and click to choose it on the remote PC, but the password must be typed in and must have the proper upper- and lowercase letters and numbers, just as we entered it when we shared the resource.

Now go back to the second PC, the wireless laptop in this case, and go through the same process in Windows Explorer and the Network Neighborhood folder. Click farther down through the network until you see the shared folder PriShare hanging off Primary PC. Right-click PriShare and select Map Network Drive. The dialog shown in Figure 5.6 will appear.

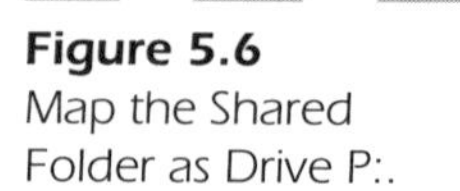

Figure 5.6
Map the Shared Folder as Drive P:.

From the Drive pull-down, select P:, check the box for “Reconnect at Logon,” and then click OK. Windows Explorer will blink and a new drive, the P: drive (for Primary PC), will appear. You will be asked for a password the first time you try to access this drive. If you always want this PC (your laptop in this case) to have access to the shared folder PriShare on the Primary PC, check “Remember Password” on the dialog that appears. Keep in mind, however, this means that *anyone* who gains access to your laptop will be able to access this shared folder, with nothing in the way but an easily defeated sign-on password. The alternative to consider, especially if you infrequently need to get to this shared folder, is to not check the Remember Password selection. Then anyone (including you) accessing PriShare will have to enter the correct password each time to access the shared folder during a session. A session is the period of a single log-on. So if you log out, shutdown, and/or reboot your laptop, the password will be forgotten and will need to be reentered.

From this point on, the shared folder will act like any other drive resource on the laptop PC. That means you can drag and drop files, open and save documents, or even delete files in the P: drive from your laptop. Also, you can run programs from this synthetic P: drive. The only difference between this networked drive and a local folder is that the actual files are stored on the Primary PC in a folder called My Shared Files.

Printer and Fax Sharing

Printer sharing is a more sophisticated task than the simple file sharing covered in the previous sections because of two factors: printers are rather slow devices, so the host PC must buffer the print data that comes from the remote PC; and the remote PC must have the proper printer drivers installed for each printer, whether locally attached or remotely shared.

With printer sharing, you can make each printer available to any PC in your local network. For those with several PCs, this allows you to put your printers where they are most convenient for you but still print to any networked printer, almost as if it were locally attached. For example, you may have two PCs that are connected to your wired network, as in Chapter 4. The most convenient arrangement may be to have a printer on each PC, perhaps the high-speed laser printer on your primary PC and a color ink-jet printer on the secondary PC. Now, with printer sharing, you can easily print on the color printer from the primary PC or on the high-speed printer from the secondary PC. If the second PC is in an area where there is little room for a printer, you can just as easily attach both printers to the primary PC and print to the color printer or the laser printer over the network.

More to the point, you can print from your wireless laptop computer without having to drag a printer or printer cable around or sneaker-net a diskette. It is access to these bulky, but useful, resources that makes sharing really powerful for wireless networks.

To set up access to a shared printer on another computer, you must complete the following steps:

1. Install the printer on the target PC
2. Configure the printer as a shared resource on the target PC
3. Add the printer as a networked printer on the remote PC
4. Print a test page to be sure the connection works properly

The first step is really a reminder. Obviously, if you want to share a printer on one PC, that printer must be properly installed on that PC. This means that you have followed the manufacturer's installation instructions and have set up the printer for local printing. You may connect the printer via a parallel printer port or a USB port, depending on the printer's capabilities. Many printers have both interfaces available. Frankly, the USB connection is much faster and more convenient. USB cables are tiny compared with parallel cables, and it is easier to use extension cables to gain a little more cable length. You can also connect multiple devices over a USB interface by using an inexpensive USB hub. The installation of a USB printer is only a little more vexing than the installation of a parallel printer because it takes special drivers, but everything else is so much better that it is worth the extra effort.

After the printer is properly installed on the target PC, you need to configure it as being shared. Open the Printers folder by selecting Start/Settings/Printers (or by opening the My Computer folder and the Printers folder). This folder will show your installed printers as icons or a list, depending on the way the folder view is set.[3] Now right-click on the printer you want to share, and a list of options will appear, as shown in Figure 5.7.

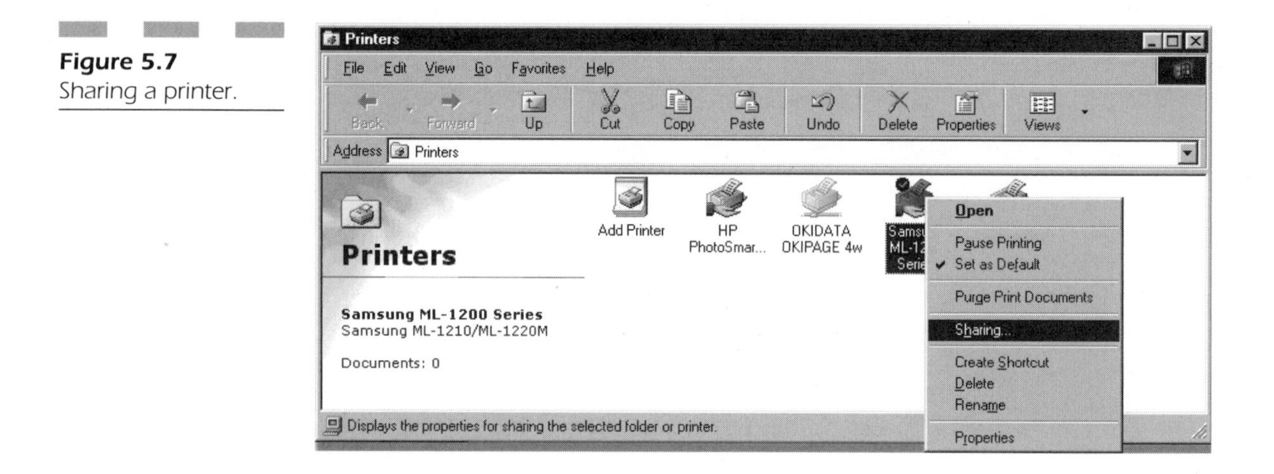

Figure 5.7
Sharing a printer.

Select "Sharing…" and the printer's Properties dialog will appear, usually with the Sharing tab displayed (Figure 5.8). From the Sharing tab, select "Shared As" and enter the share name. The best practice is to enter a name that obviously relates to the printer when viewed from another PC on the network. Remember, you will have to select this as a networked printer later. If you have only one laser printer, as in this example (Figure 5.1), you can call it Laser1. In an office with several printers, you could label them "Laser1," "Laser2," etc. On the other hand, you could share-name the printer "Laser-on-Pri" to be more descriptive. You must keep the share name at fewer than 15 characters, and very few punctuation characters can be used. Although Windows will accept a <space> character as being valid, many programs, including some from Microsoft, will refuse to recognize and print to a printer so named.

[3]You could change the view of this folder by pressing Alt + V and selecting the view you want.

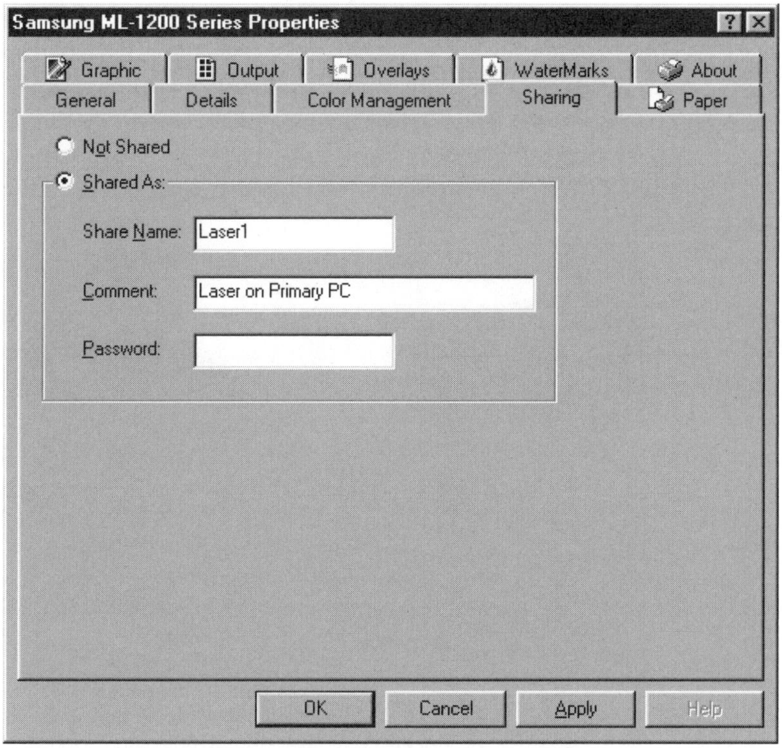

Figure 5.8
Setting a share name for your printer.

Samsung ML-1200 Series Properties
Graphic | Output | Overlays | WaterMarks | About
General | Details | Color Management | Sharing | Paper
Not Shared
Shared As:
Share Name: Laser1
Comment: Laser on Primary PC
Password:
OK | Cancel | Apply | Help

Your shared printer can now be accessed by any other PCs in your workgroup. At this point, you may repeat the process of sharing any other printer resources. Many versions of Windows are shipped with a virtual fax machine utility. The fax-sending portion of this utility is accessed as if it were a local printer. The only difference is that it "prints" in a normal facsimile format over an attached analog phone line. This is a handy capability to share with the other printers on your SOHO network, including your wireless laptop.

Installing the printer on the wireless PC. Now that you have completed the first two steps in sharing a printer over your network, it is time to go to the remote PC (the one that needs access to the networked printer) and install the shared printer there. Such printers are referred to as *network printers* in contrast to the *local printers* that are directly connected to a PC. We will assume that you will be installing this networked printer on your wireless laptop PC, although it could really be installed on any wired or wireless PC within your workgroup.

To install a networked printer on your wireless laptop, go to the Printers folder on that computer by selecting Start/Settings/Printers (or the My Computer folder and the Printers folder). Open Add Printer (double-click on the Add Printer icon) and the Add Printer wizard will appear. On Windows 9x, the Wizard first displays an informational screen, so click Next. Now choose Network Printer (Figure 5.9) and click Next.

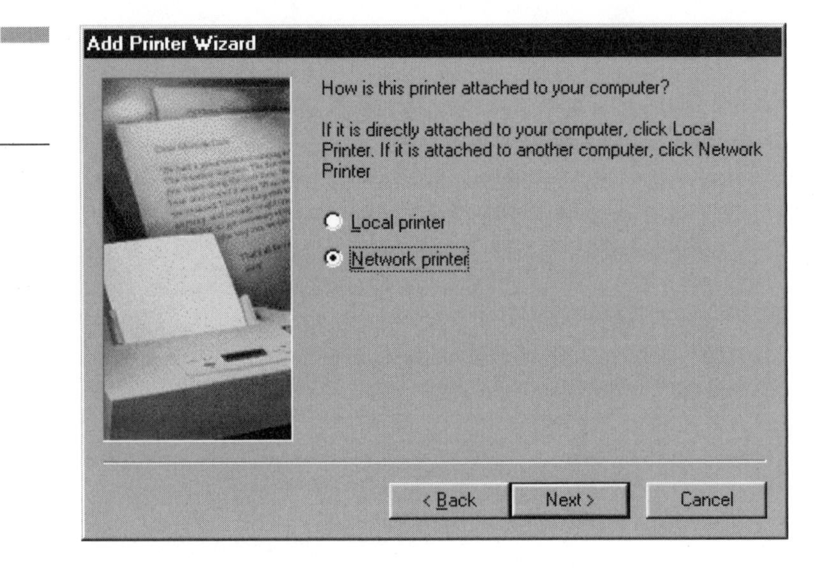

Figure 5.9 Choose network printer.

Now you will need to identify the networked printer you wish to install on your laptop PC. As was mentioned before, all the computers that are members of your Windows workgroup (as specified by the workgroup name) have unique computer names that were entered on the Identification tab in the respective Network setups. The remote computer resource can be identified by a Network Path. In the case of our Laser 1 on Pri, this would be entered "\\Pri\Laser1" (Figure 5.10).

If you would like to pick from a list, you can click Browse to display a list of printers that are shared across the network/workgroup. This list looks and works identically to that in Network Neighborhood, except that it shows the shared printers beneath the networked PC on which they are locally attached. In our example, the Primary PC name "Pri" will appear and beneath it will appear Laser1. You will also see the other shared printer devices, such as FaxPrint, if you have set them up to be shared. This list works just like the folder list in Windows Explorer. If you do not see printer objects as subitems under Pri, click the + (plus sign) to the left of the computer name to display them. In some

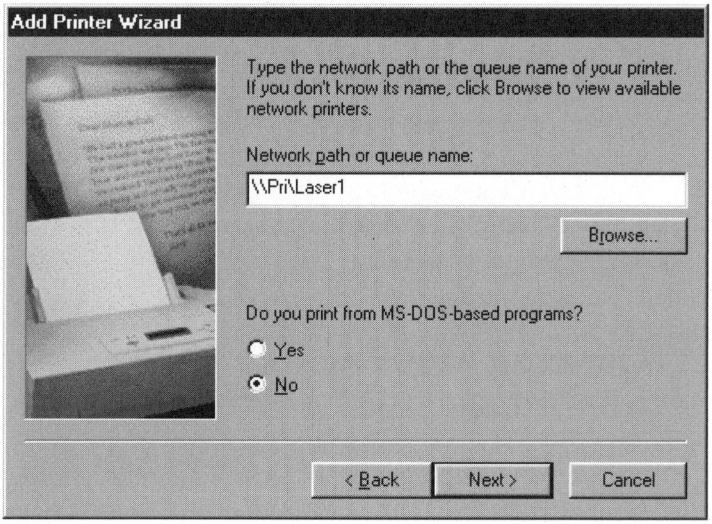

Figure 5.10 Enter your network path or click browse.

cases, you will see your own PC name and the workgroup name initially displayed as the only objects in the list. Just click the + next to the Workgroup name to expand the view to the PCs in the workgroup.

Remote printing speed versus delay. When remotely printing to a networked printer, speed is not a major concern, except in extremely intensive graphics printing, because your network runs at 10 Mbps or 100 Mbps, both of which are comparable with, or faster than, the standard parallel printer interface speed. Even the USB connection is slower than a 100-Mbps network. The only reason to move a file to the PC with the printers locally attached is to avoid the software delays inherent in converting the print job for network transmission. That conversion is what we call *processor intensive*, so the raw speed of your PC will have the most effect on your print speed over the network.

This is a good reason to place your shared printers (and other resources) on a very fast computer. Processor speed and, to a lesser extent, input/output (I/O) bus speed will have a significant effect on processing the print jobs and the print queues for your shared printers. It is quite commonplace to see a two- or three-fold speed increase from one computer generation to the next. Obviously, this can make quite a difference at both ends of the remote printing equation. More speed in your wireless laptop PC translates into faster processing of the print job, and a faster desktop PC with a shared printer results in faster execution of the printing commands as they are delivered from the print queue.

If you are connected wirelessly, you will be able to print to any networked printer in the same manner as a wired PC. However, it is more likely that you will experience longer print times because the wireless connection may not be operating at its optimum speed. If there is poor signal strength, the 802.11 network will automatically throttle back to minimize transmission errors. This means that you may be operating at a 5.5-, 2-, or even 1-Mbps data rate, which is low enough to be noticeable on a large printing task.

Copying and Backing Up to Shared Disk or CD Writer

A great use for a shared folder or CD writer is backing up your files. If you shared a folder on your Primary PC, for example, you can connect to that directory from your laptop PC as if it were local. You can very easily use drag-and-drop within Windows Explorer to save copies on the shared folder. This is a little easier if you mapped the shared folder to a drive letter on your laptop. This creates a networked virtual drive on your laptop that is linked to a real folder or drive on the target computer, such as Primary PC. You can use the networked drive as you would any other disk drive or folder on your own hard disk.

When you are dragging and dropping these files from folder to folder, it can get pretty confusing. On your own hard drive, you can do little harm because the file will always be there somewhere, just not where you thought you put it, in some cases. To demystify all this, take a look at the Tech Tip on What's All This Right-Click/Left-Click Stuff?

Tech Tip: What's All This Right-Click/Left-Click Stuff?

You can create subdirectories (folders), write files directly, and copy and delete files and folders. If you use drag-and-drop, you should use the right-click method because if you use the left mouse button to drag and drop, Windows will assume what you intend to do. And as we all know, assumptions can be very bad. For example, if you left-click and drag an ordinary data file, Windows will assume you mean to move that file. If you wanted to copy it to the mapped network drive for backup purposes, too bad! You just moved it. In contrast, if you try to left-click and drag an executable file (with the .exe extension, for example), Windows will assume that you want to create a shortcut to that file and will neither move nor copy it.

The best practice is to use the right mouse button for drag and drop. We show this operation as right-click because it is fundamentally different from left-click or simply click-and-drag. You can continue to use the left mouse button as usual when you are told to "click" or "select" something. Just don't use it for file move/copy/shortcut tasks.

When you right-click and drag a file from one folder into another, you will be greeted with a small pop-up that lets you choose the action:

Move Here
Copy Here
Create Shortcut(s) Here
Cancel

Now you are in control! For your backup tasks to the mapped virtual drive, you will always want to choose Copy. You can click (left-click, remember!) on the word Copy or you can press the letter "C." You may notice that one letter in each command is underlined. This is the shortcut command, so it is "M" for Move, "C" for Copy, and "S" for Shortcut. To cancel, you may click Cancel, press Escape, or click in an empty space over in the File Name panel. If you make this a habit, you will not only have more control over your file moves/copies, but you will have that instant whether to consider if this is really what you want to do.

One more minor point: when you use the left-click and drag method, you will see a small + or arrow attached to the cursor if Windows intends to copy (+) or create a shortcut (arrow) when you release the left mouse button. If the assumed action is what you want to happen, you can just release the button (in the correct folder, of course). If not, you should move the mouse cursor back toward the file name side of Windows Explorer until the "not" symbol, ⊘, appears on the cursor. A good place is on the vertical bar between the panels or on the original file name. Then you can let go of the left mouse button and nothing will happen.

If all else fails and you do the wrong thing, you can usually get the move or shortcut action reversed by clicking Undo from the Edit pull-down on Windows Explorer's Menu Bar. Just note what the Undo command will reverse by looking at the full description next to the word "Undo."

One way to back up your files is to move them over to a networked drive and from there transfer them to a backup tape, a CD writer, or CD-RW.[4] A common CD-R has a capacity of nearly 600 MB, which is much more data than most of us need to save. A good strategy is to back up your data files (documents, spreadsheets, financial data) to CD-ROM on a regular basis. On most systems, the process is very simple. First, you initialize a blank CD-R disk and enable it for disk emulation. In this mode, you can treat the writable media as an ordinary large-capacity disk. You can drag-and-drop files to it and copy files back from it. It acts just like any disk drive, with a few exceptions.

One difference is that the data is permanently written onto the disk by a laser diode. Any actions that later delete a file, move a file, or replace the contents of a file with more current data leave the originally written data intact, but change the CD file system's (CDFS) pointers to the disk. (This is a delight for Senate subcommittees!) What actually happens is that these actions decrease the total available file space by the amount of the deleted or original file length. On an ordinary hard disk, if you have a 100 kB file and add more information, say to make it 110 kB, the file system rewrites the entire expanded file over the original file's sectors and adds disk sectors as needed for the extra 10 kB. This is generally true with nondatabase files, even if only the end of the original file has changed. However, on the CD-R, the original file's recording has been rendered useless, as any portion of it may have been changed, so it is left intact but discarded from the disk's file allocation tables, and the entire file is written anew in the next-vacant space.

Another significant difference is that the real-actual-true directory listing (the file allocation table) can be written only once! So during the time the disk is left open for additional write operations (copy and move), the file directory lives in a state of suspended animation elsewhere on the disk. Once you are finished writing to the disk, you "close" the disk by calculating and writing the final file table in its proper place. Until then, the disk cannot be read on any "ordinary" CD drive and may not be usable on another CD-RW drive.

Most CD writers have a setting that automatically prompts the user to choose between leaving the disk as is (in directory limbo) or closing it so it can be read on any standard CD-ROM–compatible drive. You

[4]A compact disk drive in a PC is referred to as a CD drive or a CD-ROM. Originally, all CD drives were read-only. Now a type of CD drive, often called a CD-R (for recordable) is available that can write-once on special recordable media, called a CD-R disk. Newer versions can also erase and rewrite (a limited number of times) on a different media type called CD-RW (for read/write).

should always close and label the drive. The process takes a few minutes but produces a much more usable backup disk.

To share the CD-RW drive on a wired or wireless network, simply follow the same steps for sharing the C: drive. In brief, find the drive in Windows Explorer, right-click, choose "Sharing...," and give it Read/Write access, with a password if appropriate. Choose a common-sense share name, such as CD-drive or CDRW. From the remote wireless (or wired) PC, surf your Network Neighborhood, highlight the PC with the drive to be shared, and find the shared resource with the correct name (CD-drive or CDRW). Right-click on that drive and choose "Map as Network Drive." Choose an appropriate and memorable drive letter, such as R:. (Sorry, C: and D: are already taken.)

Now you have wireless access to your CD-RW drive back on that primary computer. The only thing you have to keep in mind is that you must have already placed a blank CD-R disk in the drive and initialized it for disk emulation, before you can drop files in it from the remote PC. If it is not properly loaded and initialized, you will get a drive-not-ready error. With most drag-and-drop operations, Windows will assume that you want to make a copy, because the CD-RW is considered "removable media" by the OS.

Cordless Phones and IEEE 802.11 LANs

The shared, unlicensed frequency bands of the Industry, Science, and Medicine (ISM)[5] and Unlicensed National Information Infrastructure (U-NII), as they are known in the United States, offer tremendous advantages and disadvantages. To a certain extent, they are different aspects of the same regulatory paradigm. Essentially, any device that operates in compliance with the fairly liberal standards and limits of the regulations that apply to these bands can operate with impunity.[6]

Several competing technologies have seen the 2.4- and 5.2 to 5.9-GHz bands as fair game. The frequencies are available on a interference-

---

[5]Commonly paraphrased as Industrial, Scientific, and Medical, also translated to ISM.

[6]The FCC Part 15 rules governing these devices restrict power and operating modes. Other services have *primary allocation* in the bands, which means they have priority and the right to force us to cease operation if we interfere with them.

accepted basis. As a result, quite a few technologies, including IEEE 802.11 WLANs, Bluetooth, HomeRF, and cordless[7] telephones have been developed to take advantage of these wide-open frequencies. However, all good things must come to an end, and so it is with these formerly wide-open frequencies.

The latest cordless phone technology operates at the fabulous frequency of...you guessed it...2.4 GHz, the same band as 802.11 WLANs. Of course, so do all those other technologies. Cordless telephones are only the scapegoat because they are very rapidly moving into homes and offices at the same time that large numbers of 2.4-GHz WLANs are being deployed.

As with the other services, cordless phones and WLANs are "mutually unaware." That means that they just blast out their transmissions without any idea whether there are any other services operating. Granted, they are aware of those of their ilk. That is, multiple 802.11 networks work relatively well together, although the channel frequency for the APs may have to be moved around a bit. Likewise, Bluetooth and HomeRF devices function well with their own types of devices. A way for 802.11 and Bluetooth devices to recognize each other and increase their tolerance, or rather, avoid interference is emerging.

The preponderance of interference reports seems to be with 2.4-GHz cordless phones at this point. The problem may or may not be significant because the AP or cordless base may use alternate channels to avoid each other. Many owners of both 802.11 WLANs and 2.4-GHz phones report no problems. In other cases, the problems have been of such concern that at least one major university that uses 802.11 on campus has asked students not to use 2.4 GHz phones in their residence halls.

Ultimately, all these competing services must be made to work together. Fortunately, it appears that interference is minimal in most cases. It seems likely that future revisions of most of these technologies will include a means of sensing each other and avoiding or at least minimizing interference.

Internet Phones—Voice over IP

A new voice technology called *Voice over Internet Protocol* (VoIP) is causing quite a revolution in voice telephony. This emerging technology

[7]Cordless or wireless, these telephones are intended to be relatively short-range cordless handset links for ordinary analog telephone lines.

potentially allows you to make and receive voice calls from your computer. You can literally bring your office phone to your easy chair, via your WLAN. What's more, your easy chair doesn't have to be in your office (or home office) or connected directly to your office LAN. You could be reclining half a world away, taking your phone calls, picking up your voice mail, and making outgoing calls from your office phone line.

This VoIP technology is amazing. Not only does it take phone calls and send them over any network, but it also sends the specialized signaling data with the voice. That means you can have all the specialized features from your office private branch exchange (PBX) just as if you were in the office. For example, you can have voice mail, message waiting, call transfer, call waiting, three-way calling, voice conferencing, and more.

Of course, you can use VoIP to make ordinary phone calls, too. You can make direct IP-only calls to other VoIP-equipped users right over the Internet. If you sign up with a service that accepts VoIP call traffic, you can even place calls to real, live, phones.

How VoIP Works

VoIP operates by digitizing your voice, converting it to IP packets, sending those packets over an IP network, and eventually converting the packets back to audio for the person to hear at the other end of the line. Conversely, the process is repeated in reverse, so you can hear the other person's voice.

Much of the voice digitizing technology was developed to make ordinary long-distance phone calls work better. When you make an ordinary phone call across the country or across the world, your voice is turned into a 64-kbps digital stream.[8] This digital voice channel is mixed with others and eventually transported and converted back into audio at the other end. In the middle, it has been *circuit switched* to route your voice (and the other person's voice) along a selected path. For the duration of the call, you "own" bandwidth of a 64-kbps channel for the entire physical path and the reverse path.

---

[8]Telephone engineers will insist that only 56 kbps are used for each voice channel in most cases, but the other 8 kbps are stuffed with digital 1s to maintain synchronization, so the entire 64 kbps is "consumed" by your call. This is a part of T1 technology. T1 packs 24 voice channels, each with 56/64kbps, onto a digital carrier at 1.544 Mbps.

As you can imagine, a 64-kbps channel is quite a lot for your little old voice, with its 4-kHz width. Clearly, it takes somewhat more than 4 Kbps to digitize a 4-kH signal with good fidelity. But very brilliant and innovative engineers and mathematicians have figured out how to make a very good-quality digitization and encoding at 8 to 12 kbps. This is commonly called *compressed voice*, as it uses far less than the conventional 64 kbps. At 8 kbps, you could pack eight full phone calls in each 64-kbps conventional channel's bandwidth!

At 8 kbps, you can also send that digitized signal over the Internet with very little impact on the network. That is exactly what we do with VoIP, with some additional details.

One problem with sending voice over the Internet is that sequential packets sometimes take different times to reach the same destination. That doesn't bother us when we are sending ordinary data, but it is a problem if the data packets consist of digitized voice. To operate properly, the IP connection must send the voice packets reliably, with minimal delay. Our digitizing equipment must label each packet to identify it as voice (implying a higher priority) and to identify the order of the packets. The far-end receiver, in addition to that obvious job, must resequence the received packets, if necessary, and buffer them in such a way that they can be applied to the digital-to-analog converter in a timely fashion. Otherwise, the far end would receive very choppy, distorted, voice with annoying gaps and delays.

We can also provide a path over the Internet that supports a good quality of service (QoS). Ethernet protocols already have a provision for denoting and maintaining QoS, but the proper indicators must be added to the Ethernet packets that carry VoIP packets. This is done most easily at the source.

A VoIP link can be provided by a specialized type of router or our little laptop PC. We just have to add client software to support the functions we want.

Installing VoIP Clients

To operate a VoIP capability on your wireless laptop PC, you will need to install a VoIP client. This is a fairly straightforward application that uses your audio card to "play voices" instead of music. Actually, your laptop probably does not have a separate audio card, as do many desktop computers, but has the audio functions built in. The same audio ports that play the musical "ta-da" as Windows starts can reproduce the voice from a phone call. Likewise, the microphone input on your laptop can accept your voice for digitizing and transmission.

Many VoIP clients are available in a variety of formats. These programs range from freeware, to shareware, to pay-as-you-go-ware. Special-purpose VoIP clients are also available from all of the major office PBX vendors. If you want to add this capability to your laptop, check out the following site for a listing of client software and accessory devices: www.BuildYourOwnWirelessLan.com.

Using Headsets for Better Fidelity

The speakers and microphone in your laptop computer are not optimized for VoIP use. In fact, many laptops do not have built-in microphones. You can overcome these problems with a headset designed for Internet phone use or for dictation and voice recognition. Several types are available.

The least expensive headset is a combination headset with a built-in microphone. Many variations are available, including the one shown in Figure 5.11. Models are available with over-the-head bands, around the ear clips, and in-the-ear mounts. You should pick one that is most comfortable for you within a price range that you consider affordable. There is a wide range of quality and price, as you might expect.

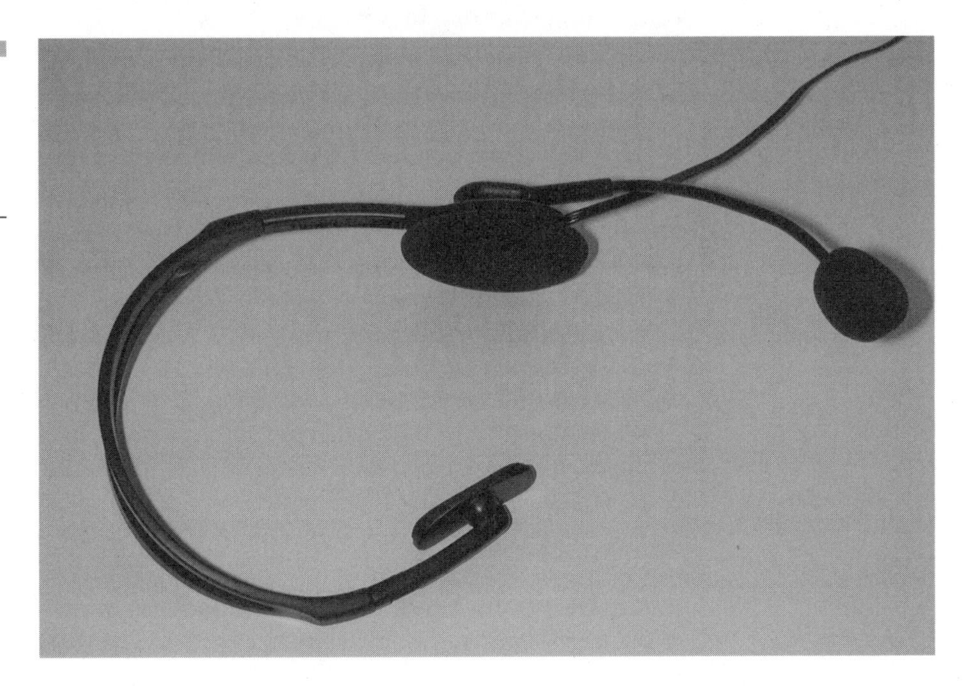

Figure 5.11 A headset/microphone combo for use with audio cards.

Many laptops have less than perfect audio systems that may cause problems with VoIP operation. If this is the case with your laptop, you should consider a headphone/microphone set that connects through the USB port. These devices digitize the voice to apply it to the USB port and may have more fidelity than a simple PC audio port. As long as the digital conversion is many times the rate that you intend to use for your Internet phone, it will deliver exceptional fidelity.

IP Phone Issues

It would not be fair to discuss IP phones and VoIP without covering some of the drawbacks. The most pervasive problem is that the Internet is a very unpredictable environment, particularly when it comes to delivering packets in a nice, neat, timely stream. If the voice packets arrive late or not at all, you will hear annoying gaps in the voice from the other end. However, the situation is improving as the total available bandwidth on Internet connections improves.

If you are accustomed to talking on a regular phone line, you unconsciously know that there is practically no delay between your voice and the other person's voice heard on your end of the line. In reality, there are tiny delays, but they are normally under a tenth of a second and virtually imperceptible in a normal conversation. However, in IP telephony, some more substantial delays are possible. A typical IP packet's transit time can be measured with a protocol called Internet Control Message Protocol (ICMP), which can send a packet to a remote location, such as a Web server or a router, and have it immediately returned. The round-trip time is measured by you, the sender. This is rather like a reflection from a radar pulse, and conveniently, the process is called a *ping*. Typical round-trip times of as much as 300 milliseconds (3/10 sec) are common when distant servers are involved. This delay (one-way) is part of what you experience on an IP call, and it can vary by more than 2:1. Also, as we have discussed, the VoIP receiver must insert an additional buffer delay, so that it can compensate for variations in packet-arrival times.

When these VoIP delays occur, you and your called party experience a problem with carrying on a normally paced conversation. You may seem to be speaking out of turn or the other person may begin speaking just as you have started to say something. If you know the other person well, it will seem that they are just rushing the conversation. If you do not know the person, a certain rudeness may be inferred, particularly if the other party does not know they are on an IP line. Once you perceive the delay problem, each of you may wait uncomfortably long before replying or interjecting, unlike your normal behavior in a normal conversation. You can learn to compensate for these problems, as many have done with push-to-talk two-way radio, but it is not what you have learned to expect from a normal phone connection.

It should be pointed out that these delay issues are essentially nonexistent on an in-house VoIP system. Many manufacturers are offering specialized IP telephones that work through a complementary IP-voice switch. These systems are scalable from a few extensions to a few thousand. Typically, they are supported by a 100-Mbps Fast Ethernet network with QoS built in. In addition, the voice is often digitized at 64 kbps, since there is so much local bandwidth available. The connection to the world of analog phone lines is done right at the IP-voice switch, so the quality is superb.

The great thing about these systems is that you can theoretically grab the IP phone off your desk, take it home, and hook it up to your Ethernet hub, thereby routing your office telephone traffic right to your home, with all the PBX features of the office. What is even cooler is that you can install an IP phone client in your wireless laptop and have all of the same features. With the advent of unified messaging, you can get your voice mail, e-mail, and faxes from your laptop, wherever you are.

Another issue to keep in mind is that there may be some difficulty in placing calls to real phones that are not connected to the Internet. You usually need to go through an intermediate point, such as your office PBX or an IP-to-analog service provider. You can subscribe to such services, but they have a cost per minute, just like ordinary long-distance charges. Fortunately, these services usually charge much less than ordinary long-distance phone charges, so you can save some money with IP.

Project: Wireless Virtual Office

This project shows you how to create a wireless virtual office with local peer-to-peer networking and VoIP. This project is suitable for a SOHO installation and telecommuting. You can connect wirelessly to all of your normal resources, including the phone, while you sit in your easy chair.

You also could use this project, with the addition of WINS resolution, to log into your server at the office and remotely connect to your office telephone system, if it is IP enabled. Perhaps you could move from the easy chair to the couch or the kitchen table for more serious work. You wouldn't have to stay at home. Lots of hotels and resorts have WLANs. If things really get tough, you could virtual office from a nice Caribbean island beach (we have several in mind, don't you?)—about two weeks on the beach in mid-February would be nice. Of course, you will have to mute the sounds of the surf when you call in on business. After all, Internet surfing is supposed to be figurative, not literal.

What You Will Need

1 DSL or cable modem and line
1 SOHO router and hub, plus cables
1 Wireless AP
1 Networked desktop PC
1 Shared printer
1 Fax modem software with phone line
1 Laptop with 802.11b W-NIC installed
1 Headset with microphone

Project: Wireless Virtual Office

Project Diagram

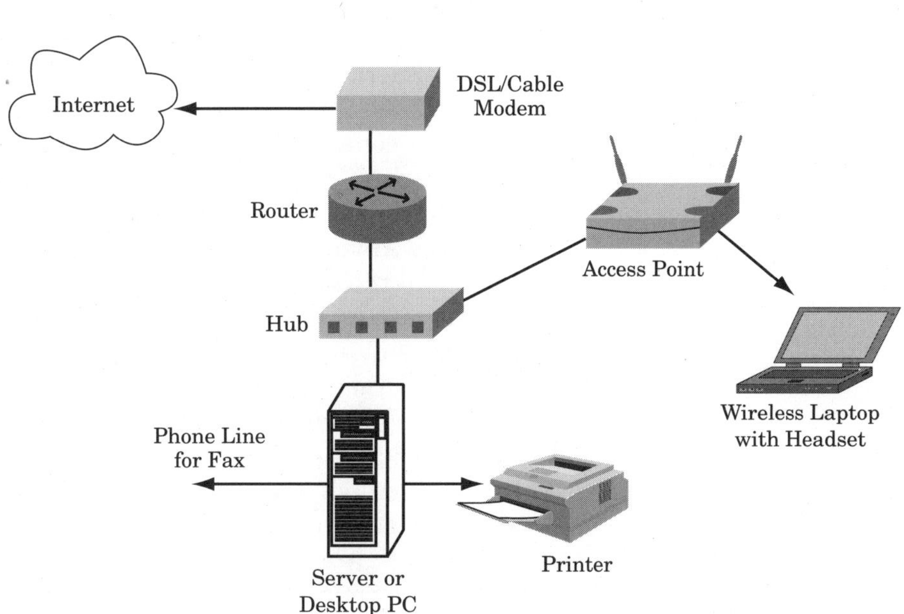

Project Description

This project recaps many of the virtual office capabilities that we covered in this chapter. With the knowledge gained in the projects in Chapters 3 and 4, you will assemble and configure the wireless and wired network components needed to have a virtual office on your wireless laptop computer. We also provide IP telephone capability and cover some of the issues for e-mail applications.

We start by creating the basic SOHO network with wireless components. Then we will add printer and fax connections and cover a few points about e-mail clients. Finally, we add an IP phone, and your virtual office will be complete.

Project: Wireless Virtual Office

Step-by-Step
Step 1: Install the Wireless Laptop and AP

This step and the next one repeat the project in Chapter 3. Consequently, we merely point out the major activities here and refer you to the previous project for details.

The purpose of this first step is to create the wireless link between the AP and laptop computer. Most wireless starter kits have an AP bridge and one or more W-NICs. Or you may choose to buy them separately. If the defaults are compatible, you should be able to set up the WLAN between the AP and the laptop with very little configuration.

Connect power to the AP first, so it will boot up and load its default settings. Next, install the W-NIC, as covered in the previous project, and start up the link test. If you wish, you can also load the AP manager utility and use it for the link test. If you have a good link, you can proceed to the next step *after* you change the default network name and optionally set up encryption. "Better safe than sorry," the expression goes. Plus you will likely forget to do this later, if you do not do it now. Be sure to write the password down in the AP manual, just in case.

You can also configure the network for TCP/IP and Windows networking at this time. Set up TCP/IP for automatic IP and set your workgroup name (on the Identification Tab) to something other than the default. You do not need to provide File and Printer Sharing on this laptop because that is a function of your wired primary PC. All these details are in the prior two projects.

Step 2: Connect the DSL/Cable Modem and SOHO Router

As we covered in the Chapter 3 and 4 projects, you can now connect the DSL/cable modem to the high-speed line, and connect the SOHO router to the DSL/cable modem. Make sure you have the proper power, line, and link lights on each device. If you power on and connect the modem before powering on the router, the router will be able to pick up a valid IP address from the ISP.

Connect the AP to the router, preferably through the hub, so you can easily add your primary PC in the next step.

Now you can boot up the laptop or renew the DHCP to gain a valid local address from the router. This should allow you to connect to the Internet from your Web browser. If you have a problem, start troubleshooting by connecting to the router's configuration screen (this should be an local address such as 192.168.1.254). Be certain that the router is set to act as a DHCP server (to provide your local PCs with IP addresses) and a DHCP client (to grab a valid IP address from your ISP). Check the DNS settings, too. In general, you will have to manually set them on your router to the DNS addresses of your ISP (two addresses, a primary and an alternate, are customary).

While you are in the configuration screen, you should change the default password for the router. This is a major security hole that needs to be plugged. Now write the password down in the router manual because it is critical that you have it to make future changes or to monitor your connection.

Step 3: Connect the Desktop PC and Hub

The next item of business is to connect your primary desktop PC, or server, to the Ethernet hub. This will make it available for resource sharing to your wireless laptop and coincidentally will make it accessible to the Internet. This setup was also covered in a previous project, in Chapter 4, so we refer you to there for the details. If you have followed the projects in order, you already have completed this and most of the next step.

You should confirm that you have shared the proper directories, printers, and fax modems (the virtual kind) for remote operation from your wireless laptop PC. Recall that you may (and should) password-protect access to these shared resources, and that you will have to enter the appropriate password when you first access one of them from your laptop. If you check the Remember Password box on the password pop-up, you generally won't have to do this again. Also, if these are mapped resources, such as a disk drive or a folder, you should check Reconnect at Logon.

Keep in mind that, if your wireless connection is not operating or your primary PC is not booted up and properly connected, these auto-connecting resources will time out and present you with an error message. The message, which will be on your laptop, will say that the resource is "not responding" and give you the opportunity to acknowledge the message. Be sure that the option to reconnect the resource is checked properly. If not, the connection (and mapping) will be broken and you will have to manually reconfigure it. Also, on Win 9x machines, the resource will not automatically reattach when it becomes available, without a reboot.

Step 4: Configure Peer-to-Peer Networking to Resources

If you have not already done so in the previous project, you should configure the resources on your primary desktop PC for sharing and configure the network and workgroup settings. You will need to set up File and Printer Sharing on this PC.

At this point, you should be able to "see" the other computer(s) on your peer network by opening Network Neighborhood. Alternatively, you can look down the Folders list in Windows Explorer and open Network Neighborhood that way. Sometimes peer networking is a little frustrating to set up. Remember that you need the Client for Windows Networking and generally will be running NetBIOS over NetBEUI protocol, unless you have the NetBIOS over TCP/IP enabled.

If your primary PC is running Windows server software, which means you have established an NT domain, you will need to log onto that domain rather than use peer networking. If so, you should add the Microsoft Family Logon client to the installed components in your Network properties and set it for the Primary Network Logon. You can also reach shared resources by setting a domain logon for your Client for Microsoft Networking properties.

Step 5: Share the Printer and Fax Modem

All the devices you might want to access from your wireless laptop should be enabled for sharing and installed as network devices on your laptop. The details for this step were described earlier in this chapter and in the project in Chapter 4.

Many computers come with "light" versions of a fax emulation client. This may be an included feature, if your PC came pre-setup with OS and applications, or it may be an additional resource available on your setup disk, perhaps one that was not initially loaded to conserve disk space. In these days of massive hard disks, there is little reason not to do a "full" installation, but the installer applets make it easy to choose "typical installation," which often leaves out very useful helper applications, such as the fax software.

Fax emulation is an application that lets your computer send and receive faxes over a phone line, just as if it were a stand-alone fax machine. Of course, you won't have a scanner, unless you have added one. The fax emulation comes with two components, a fax sender and a fax receiver. The purpose of the fax sender is to allow you to transmit a printable document to a distant fax machine, using the built-in 56 kbps analog modem (now on most computers) and a regular phone line. (If you bought an after-market modem card, it probably set up the fax emulation automatically during installation.) The fax sender pretends to be a printer, so your word processor or spreadsheet program can print to a normal-looking installed printer. Virtually anything that can be printed in Windows can be faxed.

The easiest way to determine whether your computer has fax emulation is to look in the Printers folder. You can open My Computer and double click the Printers folder, or you can select Start/Settings/Printers to open the Printers folder. The fax client will show up as an ordinary printer with a telling name, such as "Brand X Fax Starter Edition." Just look in your computer's documentation, if you do not find the fax printer installed. You also may obtain retail versions of faxing software, which is a really good idea if you plan to do a lot of fax emulation.

You can receive faxes on your fax modem with fax emulation software, too. The fax receiver can be set to auto-answer and to put the received fax in a particular folder. If you are a clever WLAN user, you can set this to be a shared folder so you can view your received faxes from your easy chair.

Some printers can do double or triple duty as a combination printer, fax, and scanner. These printers have their own software for faxing and receiving, which you will probably want to use in place of the internal fax modem. The scanner is handy for scanning documents to be faxed and for copying documents (which is really a quadruple use, isn't it?). The only drawback to these combo fax/printers is that the driver program must continually monitor the fax receiver for an incoming document (or the scanner for a scanned one). This process, which runs in the background, can become really annoyed if you shut the printer off or do not always connect the printer cable. As a matter of fact, the monitoring program can virtually bring a computer to a standstill, with jerky, stuttering cursor movements and long delays in response to commands, if it cannot find the printer. Keep this in mind, especially with laptops, because the monitoring software often is automatically loaded each time you boot up. Frankly, the combo-printer manufacturers should put a harness on this bothersome behavior or give the user some better tools for turning it off when it is not needed.

Step 6: Install VoIP

Once you have TCP/IP networking set up, you have the underlying component for VoIP in place. All you need to do is to add the application that converts your voice to a stream of IP packets and converts the received IP-voice packets back to audio. Virtually all laptop computers come with built-in audio processors and speakers. Some actually have some very fancy audio systems from manufacturers well known to audiophiles. Most newer laptop computers also have a built-in microphone, often located along the top edge of the tilt-up display screen. Laptops also normally have jacks for an external microphone and external speakers.

You can download some excellent IP-voice clients from the Internet. IP-voice programs are also available in retail computer stores. Many add-on modem boards now come with this capability. Also, all of the IP-voice switch manufacturers have software clients available to implement their special PBX-like phone features over your laptop. This information is changing and expanding rapidly. Rather than try to give a partial list here that would soon be out of date, look at the following site for information and links: www.BuildYourOwnWirelessLan.com.

The IP-voice software loads just like an ordinary application program, except that you may have to supply information to assist with placing and receiving calls. If you are using a general-purpose VoIP client, you can connect to another user in one of two ways. One way is by knowing the IP address of the other VoIP user and "ringing" that user. If their VoIP client is set to automatically answer an Internet call, they will hear a ring and have a message box pop up with your identification. When they accept the call, you can both talk.

Remember that if you are using your built-in speakers and microphone, IP-voice will act sort of like a speakerphone with a fairly long response delay. It will usually be a *half-duplex* conversation, where only one of you can talk at a time, much like a regular speakerphone. However, this speakerphone is using microphone and speaker components that were not designed for this purpose, and you must allow for the delay of the Internet and the audio-to-IP encoding.

Some programs may allow *full duplex* operation, more or less like a normal telephone. The truth is that we really don't attempt to talk and listen at the same time, unless we are interrupting or acknowledging (or talking to our kids).

If you are using one of the IP-voice clients from an IP-PBX manufacturer, you will need to follow specific configuration procedures after you install the software. The system administrator will give you instructions for the client setup. Typically, you would use a login and authentication process to gain access to your host PBX. The company would not want just anyone to be able to attach as a phone user because it would be a security risk.

It is now time to install your headset/microphone. Attach it to the laptop, using the mike-in and speaker-out jacks. If you obtained a USB headset/microphone, you will need to install it, according to the manufacturer's instructions. USB devices are very easy to install on all Plug-n-Play™ systems. All you do is connect the USB cable between the device and the laptop's USB port. You will be prompted for the driver disk and led through the rest of the installation. In some cases, you also need to install applications software after the drivers are loaded. There are some very nice voice-recognition applications that come with a mike/headset that will work for both purposes. You could use it for VoIP and for dictation of your letters and documents. Imagine that: you are free not only of network cables but also of keyboards.

Project Conclusion: Virtual Office from Your Easy Chair

Now you are truly ready to virtual office, with your WLAN, your laptop computer, access to Internet, files, printers, faxes, and phones. Now may be the time to get an extra battery for that laptop.

You can run most applications from the wireless laptop. For example, you can compose letters, create proposals, edit spreadsheets, and balance your personal accounts. You can also do all of your customary Internet tasks, including web browsing and e-mail. You can make purchases, pay bills, trade stocks, and book travel. At the same time, all of your normal printer and file resources are at hand and available for instant wireless use.

E-mail clients, such as Outlook™, Netscape Messenger™, and Eudora™ will operate just fine. If you normally keep your e-mail correspondence on another computer, such as your primary PC, you will want to change the settings in these e-mail clients so they just "read" but don't download your e-mail. The setting will be "leave on server."

Project: Wireless Virtual Office

You can also access Web-based e-mail through your browser. If you keep an e-mail account with one of the Web portals, you can handle that mail just as you always have.

But, don't be a Web-head, glued to your virtual office in that easy chair. Get out...see the world. Try the couch, or the table, or even the deck. Hey, you could recall your college days (if you are not still there) by sitting out on the roof...except that this time you would have a wireless connection to the world. All things considered, a wireless virtual office will give you the freedom to pick from your choice of locations.

WLAN Security

Chapter Highlights

- Broadcasting Your Data
- Built-in WLAN Security
- Encryption and VPNs
- NAT Firewalls
- Personal Firewalls

Security is a worrisome subject for computer systems. We all have seen the results of security breaches in the press and in the movies. The reality is probably a good deal worse than any of these reports or depictions, although probably not as magical as the movies. The fact is, most modern computer OSs were not designed with real, solid security in mind. Some would say that the state of security on these systems represents organized, reckless abandon rather than well-engineered, secure protection.

There has certainly been some purposeful misdesign of the popular OSs. That is because, in many ways, the design holes were deliberate. In some cases, security was intentionally compromised to enable powerful software features that would crush (and indeed have crushed) the competition. Whatever the purpose, the result has been very successful. We now have some of the most powerful and most easily penetrated software systems that could have been imagined, even by the most forward-looking science fiction writer or spy-versus-spy novelist.

Regardless of where the blame is placed, we must be ever mindful of the security flaws of our systems and must attempt to defend our computers and our documents against the exploits of those who would do us harm. The additional dimension of wireless networking only adds to the complexity of the problem. This chapter gives you some weapons for that defense, particularly with regard to wireless networking.

Broadcasting Your Data

Whatever your level of understanding of data transmission and system security, the use of a WLAN throws some new variables into the equation. We used several analogies earlier in this book to help you better understand wireless technology. One of those was the light bulb. We said that the invisible wireless radio waves propagated rather like the light from a bulb, except that they could move through solid[1] objects with impunity. That is basically true. Also true is that the wireless signals dim as they move farther from the source, in this case, a WLAN device, rather than a light bulb.

As you approach a wireless AP or W-NIC from a distance, the WLAN signals get stronger and stronger. At some point of approach, you can easily pick them out of the air with compatible receivers. That is what we do when we put another WLAN station, say a laptop computer with

[1]Nonmetallic solid objects, at least.

a W-NIC, within the signal range of another station, say an AP. This is illustrated in Figure 6.1.

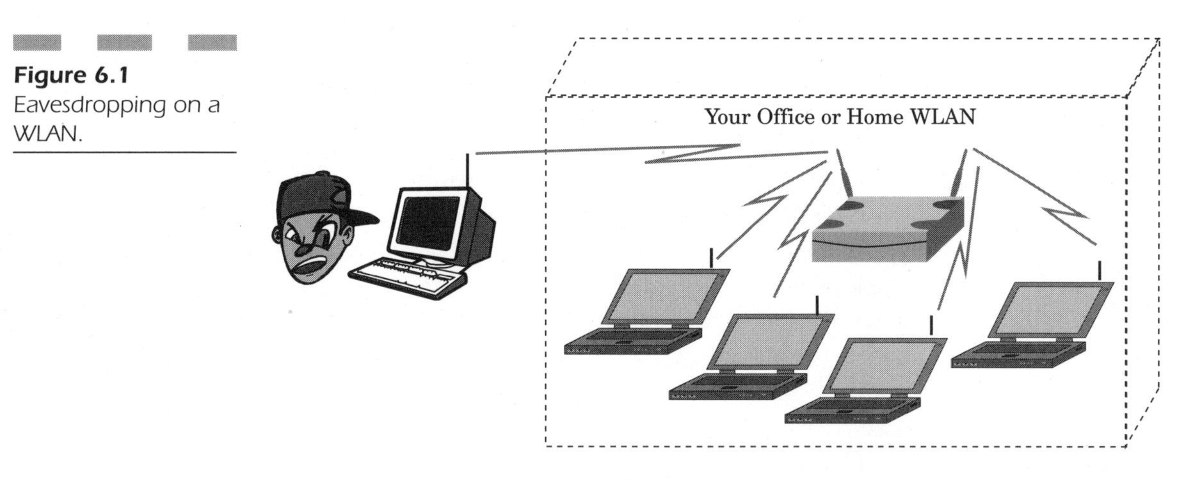

Figure 6.1 Eavesdropping on a WLAN.

In fact, if you wanted to eavesdrop, you could use a high-gain antenna, perhaps an amplifier, and a standard W-NIC–equipped PC to listen in on the WLAN communications in a building or home. The default configurations of most WLAN products turn off virtually all built-in security, so the job is almost trivial. Even the Network Name (SSID) defaults of most systems provide little protection from intruders.

Being able to access a WLAN from a moderate distance might be an advantage, if you were next door to someone with a WLAN and mutually agreed to share a high-speed Internet connection. But what if that person had no idea you were using their bandwidth for free? The same thing could happen in publicly accessible wireless Internet locations, such as a coffee shop, bookstore, or hotel. You might be eating a bagel in the store next door, rather than sipping a latté. Because the coffee shop was paying the bill for the Internet connection and the WLAN, presumably to attract customers, they inadvertently would be giving the service away, without selling their own products. Of course, you might have to be registered to use the particular Internet provider, but if there were no variable cost, other than a flat monthly fee, how would the coffee shop make sure that only their bona fide customers had access to their system?

The radio waves that carry the wireless signal do not recognize walls, much less property lines. It is as if the WLAN signal were a broadcast station that anyone with the proper receiver could tune in to. The only difference is that commercial broadcast stations operate with high-power transmitters and well-placed antennas. The potential eavesdropping W-NIC can be anywhere within range of your WLAN, which can be quite a distance with highly directional antennas.

Wireless Signal Range

You might be wondering, What is the maximum range of a 2.4-GHz or 5.x-GHz WLAN signal? The answer depends in part on the link speed you achieve. As you recall, one of the features of 802.11 wireless networking is that the transfer speed of the data payload is notched downward as the signal gets weaker. Table 6.1 shows the 2.4-GHz outdoor speed-versus-range specifications of one popular manufacturer.

TABLE 6.1

A Manufacturer's 2.4-Ghz Outdoor Speed versus Range Specifications

Outdoor Range*	Speed
150 m (500 feet)	11 Mbps
270 m (880 feet)	5.5 Mbps
400 m (1300 feet)	2 Mbps
460 m (1500 feet)	1 Mbps

**Distances are best case and the conversions between meters and feet are approximate.*

Keep in mind that the outdoor ranges do not include building structures, such as walls or floors. However, the ranges are also with the relatively puny built-in antennas for the W-NIC and AP.

It is quite feasible to extend the range of a WLAN over great distances with the use of highly directional dish antennas. With these antennas, you can actually get a usable WLAN link at as far as 15 miles (at up to 2 Mbps, possibly more). This distance assumes line of sight between the two sites, 50-foot towers, and dish antennas on both ends. But think of it…15 MILES!

If someone were trying to eavesdrop on a WLAN, they certainly would use a high-gain antenna at their end. Now, you would not expect to get 15 miles, and they would have to be line-of-sight, but they could probably be quite some distance from your home or office and still get a usable signal at whatever data rate you were operating your WLAN.

Evaluating the Risk—Indoor Range

What can be done to protect the integrity of your wireless network? With all these facts about wireless snooping, you may be somewhat concerned about the integrity of your own wireless network. Don't be too alarmed. The truth of the matter is, there are practicalities to wireless snooping that make it much more difficult than most security issues on the Internet.

The greatest protection for your WLAN is plain old physics. For someone to be able to snoop, they have to be quite close to your network. The signals decrease quite rapidly with distance and transmitted power is very low. In most instances, you will use your network inside a house or a building, and you will be restricted to the inside range specifications. Using the typical manufacturer's 2.4-GHz indoor guidelines, look at the indoor range in Table 6.2.

TABLE 6.2

A Manufacturer's 2.4-Ghz Indoor Speed-versus-Range Specifications

Indoor Range*	Speed
30 m (100 feet)	11 Mbps
50 m (160 feet)	5.5 Mbps
70 m (220 feet)	2 Mbps
90 m (280 feet)	1 Mbps

**Distances are best case and the conversions between meters and feet are approximate.*

That looks a little better. These typical ranges mean that an eavesdropper would probably have to be fairly close to your home or office to pick up a signal, although these ranges can be extended somewhat with high-gain antennas and amplifiers.

Another point is that the distance over which the signal is usable depends on the data rate that your wireless link is actually operating at. For example, if your network were operating at the full 11-Mbps rate or higher (in the case of advance standards, 22 to 54 Mbps), you would be only vulnerable to 30 m from the AP (plus whatever advantage came from the snooper's antennas). In addition, any walls, window screens, tinting, trees, or metal objects would reduce the distance even further. This is one instance in which the relatively short range of WLANs is an advantage.

On many APs you can restrict the operating speeds to any of the full or fallback values. Allowing only the higher speeds would limit the range from which an intruder could access your network.

There are also some built-in security measures that act in your favor in IEEE 802.11 wireless networks, and there are more steps you can take to make wireless eavesdropping very difficult indeed, as we will see in the following sections.

Built-in WLAN Security

WLAN technology was designed with a nominal amount of privacy built in. Perhaps this is a better term for what the marketplace and the press has confused with security. As a matter of fact, the built-in privacy feature is called wired-equivalent privacy (WEP). The idea was to give the WLAN user a measure of privacy that would be equivalent to the protection against eavesdropping that is afforded on a conventional wired LAN.

The original WEP offering was a minimal 40-bit level of encryption. This was intended to offer a moderately robust level of security at the time it was proposed. In addition, the 40-bit limit met the export restrictions at that time, which was important to wireless manufacturers. As we all know, 40-bit encryption is no longer considered particularly secure, and the export restrictions have been revised to allow much more secure levels of encryption. However, WEP was intended only to offer privacy, not security. Even wired LANs emit RF signals that may be picked up and decoded by sensitive receivers. If you attach directly to any shared LAN, you virtually own that network's data flow. Switched LANs only make the job a little more difficult.

Responding to the relatively low level of protection offered by 40-bit WEP, some of the WLAN manufacturers quickly responded with premium WEP encryption options of 64 and 128 bits. Because each added bit represents a power of 2, a move from 40 to 64 bits is a major increase in protection, and a 128-bit encryption key is a very secure mode.[2] For that matter, until recently, a 64-bit key was considered quite adequate, but processor speeds and multiprocessor computing have changed that.

[2]The 128-bit keys are not double the protection of a 64-bit key. The difficulty of breaking a key goes up as a power of 2. Thus, a 65-bit key is double the protection of a 64-bit key. A 128-bit key is 2^{64} times as difficult to break as a 64-bit key. That is 18×10^{18} (18 followed by a string of 18 zeros) times more difficult!

Security Holes in WEP

A fairly serious flaw exists in standard WEP encryption that lets an eavesdropper have access to the encryption key. The full explanation is too technical to present here, but basically WEP encryption works by sharing a set of four encryption keys between the AP and a W-NIC. At the time that the W-NIC first connects to the AP, the key is exchanged in a manner that allows a third party to observe (wireless-wise) and obtain the key. From that point on, the unauthorized station can monitor and decode all transmissions between the AP and the authorized W-NIC during that session. This flaw can compromises the security provided by the high bit-count encryption keys unless other steps are taken to secure the key exchange.

A lot has been made of this flaw in the press. The standard defense is that WEP is an optional feature that was never designed for a significant amount of security. The manufacturers plead that they are only following standards in place. Nonetheless, any encryption system that can be quickly and easily cracked through a trivial exploit causes great concern.

The reality is, only a handful of potential threats exist for the casual SOHO user, and there is probably not much to worry about. Unlike some of the snooping and break-in gambits that are possible over the Internet, from anyone anywhere in the world, the WLAN signals are physically constrained. Someone would practically have to be under surveillance by law enforcement or corporate spies to have a real security risk. In such cases, there is a lot more to be concerned about than a few stray wavelengths!

WEP key exchange was designed to be quick and easy, especially to the user. The administrator simply types a WEP pass phrase into the AP's management utility, and four keys, of the appropriate bit lengths, are automatically generated. On connection for a session, the W-NIC requests the key, which is then sent out by the AP. Communication between the AP and W-NIC from that point on is encrypted to 40, 64, or 128 bits and is fairly, very, or extremely secure, respectively—unless you have the key.

There are two problems in this scheme: the key exchange mechanism is primitive and the four keys stay the same, unless you manually change them. Because the keys are sent out without a truly secure mechanism, an interloper can observe the key exchange and gain access to the session to its conclusion (Figure 6.2). Over time, all four keys may be discovered. A clever miscreant could also trace and record all the encrypted traffic until all four keys are known and then go back and decode all the data.

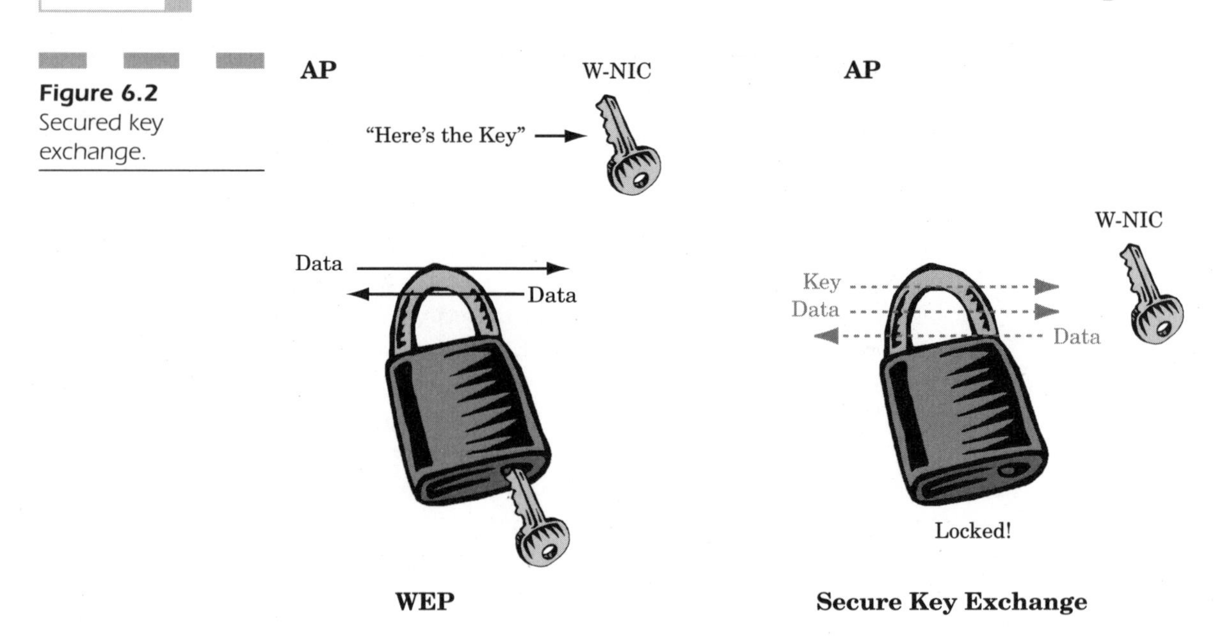

Figure 6.2 Secured key exchange.

There are a several possible solutions to this issue. One could simply change the pass phrase often. Once a day would be the minimum. This job is easily done by the administrator with the AP management utility. Unfortunately, for you, the SOHO user building your own WLAN, the administrator is you. This is a little bother that you probably would not do on a regular basis.

Another technique is to use one of the WEP updates that some manufacturers have started to deploy. These vendors increase the frequency of key exchange to as often as every few minutes. Some of the manufacturers have even added secure Diffie-Helman key exchange. Several vendors offer per-user/per-session keys tied to an authentication server (see the next section). Unfortunately, the standards are not complete on these updates, and you might get locked into a single vendor for future additions to your wireless network, if you use their improved WEP enhancements.

Newer WLAN devices very likely will have greatly improved methods of key exchange and authorization. As these new methods are standardized and approved, it may be possible to retrofit these stronger security features into the older W-NICs. Most of the W-NICs now being shipped have the ability to load extensive changes to their feature software (firmware) by means of *flash memory*. However, some of the older designs, which have been shipped and may continue to be shipped by the wireless manufacturers, may not be able to accommodate the new code and will always have these security flaws. Your only choice in this event is to replace the older cards with new ones, if you want or need to use the enhanced security features.

Authentication

Another aspect of WLAN security is the authentication of the potential wireless user. Authentication ascertains the identity of the station (the wireless node) and determines whether that station has authorization to make a wireless connection to the AP. The user may be involved in providing an additional password in the authentication process, so that the identities of user and station are assured. IEEE 802.11 has a rudimentary authentication feature built in. The AP is given an identifier, the SSID. The SSID is often called the Network Name, which is probably a more apt name. The purpose of this ID is two-fold. First, if you are in a strange Network Neighborhood (sorry, but it's appropriate) and looking for an AP, you can have your W-NIC scan for the available APs. Those within range that are open for access will respond to the scan with their SSID. Quite often the AP administrator uses a fairly descriptive name, word, or phrase, which lends itself to the term *network name*.

Theoretically, you cannot connect to a particular AP unless your W-NIC matches that AP's Network Name. So, seeing that Bob's network and Ray's network are available, Bob would naturally pick his own network to link to with his wireless connection. Now, it would be a problem if Bob decided to choose Ray's network, if he was not authorized to use it.

There are several reasons that SSID does not provide much security. One is the problem we just saw. If APs are set to respond to a scan, anyone trying to connect can easily see the available network names, set their SSID to match, and connect. A cure for this problem is to set the AP so that it will not respond to a network scan. You can then require that any connecting W-NIC have prior knowledge of the correct network name. Of course, our intruder might just listen (wirelessly) for the SSID to be passed. Finally, many APs periodically broadcast their SSID, without waiting for a scan, so any station in range will know they are available.

One answer to this access problem is to provide an independent means for authentication of a W-NIC before it allows access to the AP. A variety of methods are available for authentication. Here is a short list:

- Access control lists based on MAC addresses
- RADIUS user authentication
- Manual or electronic secure key
- Extensible Authentication Protocol (EAP; RFC 2284)

Access control lists require each AP to keep a list of the physical MAC addresses of every wireless device with access privileges. This is very difficult, if not impossible, if there are hundreds or thousands of authorized stations. In addition, it is possible to fake (spoof) a MAC address, if one knows the MAC address of an authorized user or observes the address during a wireless session. Another method is to authenticate through a RADIUS server. A RADIUS server provides a secure means of authenticating users and is a widely used standard. RADIUS also works with EAP and IEEE 802.1X.

Secure key devices are available that can allow the user to enter a unique key that is encrypted via a time-based algorithm. Ordinary digital clock technology is stable enough to allow a synchronized key (a passcode) to be generated once every 10 seconds or so. This passcode is entered by the user and compared with the time-based passcode sequence at a central location. A user password is also added, so that the secure key cannot be used when it is out of the user's control. If the passcode and password are verified, the user is authenticated and allowed access. These devices are also available in a fully automatic electronic form that is simply plugged into the user's PC and queried by the server as needed.

Combined with secure WEP encryption and secure key exchange, authentication can be a very effective means of limiting access to a wireless network. Through encryption, even a wireless trespasser cannot interpret the encoded transmissions.

Encryption—Optional Wireless Security

Beyond the fairly simple encryption of WEP are many optional security platforms that offer the promise of essential unbreakable cryptography. Although a rigorous discussion of cryptography is beyond the scope of this book, we can present some of the basic concepts. Frankly, modern encryption is like so many technical advances in our society. You really don't have to know much about how it works...just how to work it!

In this section, we cover some of the basic concepts of data encryption. We also show you how a special type of cryptography can be used to produce your own virtual private network (VPN) right in the middle of the information highway.

Public and Private Key Encryption

One of the truly unique types of encryption that has arisen in the past few decades is a concept that allows us to decipher encoded data messages without ever knowing the encryption key. The concept is called *public key encryption* and is the basis for most of the current secure data exchanges. This is what runs electronic commerce (e-commerce), authentication, and secure e-mail.

We say that an encrypted data stream is *secure*. The idea is that secure data gives us total privacy, which yields security in communication. In ancient times, ingenious ways of disguising the contents of messages were employed to send secret orders, plans, and other information across borders between nations. Frequently, a code was used that only the sender and the person receiving knew. The messenger could safely carry the message with no possibility that it could be compromised by discovery of a written communication or by torture to extract a verbal message.

Fortunately, we don't have to take a tortuous path to encode or extract our secure messages. The powerful high-speed processors in our systems do all the work for us. We use ingenious mathematical algorithms to encode data with standardized methods, so we can later decode the scrambled packets and recover the original data.

Encryption uses a mathematical formula, called an algorithm, to encode data using a key. The key is typically a string of bits that is 40, 56, 64, 128, or longer. There are two major methods of encryption, *symmetrical* and *asymmetrical*. Symmetrical encryption standards use the same key to encode and decode the data. Both sender and recipient must have exactly the same key for the message to be successfully conveyed.

Asymmetrical encryption uses two keys that are mathematically related, a public key and a private key. This style of encryption is also called *public key infrastructure* (PKI). The sender encodes the data with the private key, and the recipient decodes the data with the public key. Essentially, what happens is that the intended recipient generates a public/private key pair, and publishes the public key (or simply sends it to the would-be sender). The beauty of this method is that any data encoded with the public key can be decoded *only* with the private key, which the intended recipient keeps secret. The sender then encodes a data transmission with the recipient's public key (now the data is totally secure) and sends it to the recipient, where it is easily and uniquely decoded. This breathtaking (and mind-boggling) process is illustrated in Figure 6.3.

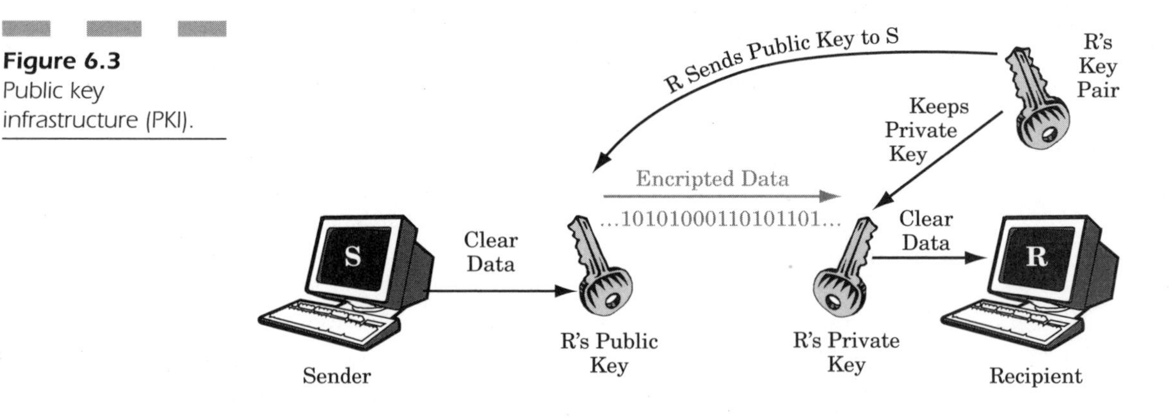

Figure 6.3 Public key infrastructure (PKI).

The tough part of public key encryption is how users can be sure that they have a legitimate public key. An elaborate system exists for distributing public keys through a *certificate authority*. Theoretically, you could supply a public key to a sender by sending it via e-mail. However, how can senders be assured that they had received the proper key? Another way of public key distribution is to send it on diskette or CD. This is exactly the way that most web browsers get their initial set of public keys. With these certificates, the browser can connect securely to a Web site with secure socket layer (SSL).

Data streams are encoded in blocks of data using the encryption key and the encryption algorithm. For example, a 56-bit key may be used to encode data in 64-bit blocks: the greater the key length, the greater the relative security of the encoded data block.

Some of the more common encryption algorithms are Data Encryption Standard (DES), RC4, Pretty Good Privacy (PGP), and Blowfish. DES is a 56-bit standard that has been somewhat eclipsed by modern processor speed. However, it does represent more than 70 quadrillion possible keys, so it would take your average supercomputer at least a couple of days to break. Don't worry, though: there is 3DES, or triple-DES. Triple-DES runs the DES encryption three times. This makes it much more secure, but unfortunately takes three times as long. However, it can be adapted easily to devices that implement DES in hardware. That should hold you for a while.

Some of the standards, including the widely available PGP algorithm, can use up to 512-bit keys. Unfortunately, more bits means more proportionally encoding/decoding time, so we usually use the shortest key that is considered really secure.

A new standard algorithm, the Advanced Encryption Standard (AES), is being developed. This new encryption method aims to give us a secure medium for the next three to four decades. As with everything in computing, time makes all things obsolete. Breaking encryption algorithms by a brute-force attack is done by going through all the possible keys using the target algorithm. It has been estimated that, if you had a computer that could break the DES algorithm in 1 second, it would still take about 70 trillion years to break AES. That's fairly secure, even in galactic terms.

Virtual Private Networks (VPN)

A specialized use of encryption is the virtual private network (VPN). One of the drawbacks of file encryption and transmission is that it is generally a manual process. With the exception of the SSL encryption done by Web browsers, most programs do not support any type of encryption. Only a few e-mail programs support the sending of encrypted documents, and even then the user must take a manual action to encode the e-mail.

A VPN allows you to create a private network between two points that automatically encrypts all transmissions. For some time, it has been possible to lease private circuits for data transmission between sites. For example, frame-relay circuits are a type of private circuit that is relatively secure, although not totally so. However, over the Internet your data goes through a variety of pathways, switches, routers, and providers. Theoretically, anyone along the path can pick up your data and look at it without your knowledge.

What a VPN does is to create a *virtual* network over the public Internet. The virtual network is made private by encrypting all transmissions. Anyone who intercepts your communication will capture coded data that is useless. This process is illustrated in Figure 6.4. It makes no difference whether your VPN node is on a wired LAN or on a WLAN. The process works the same way.

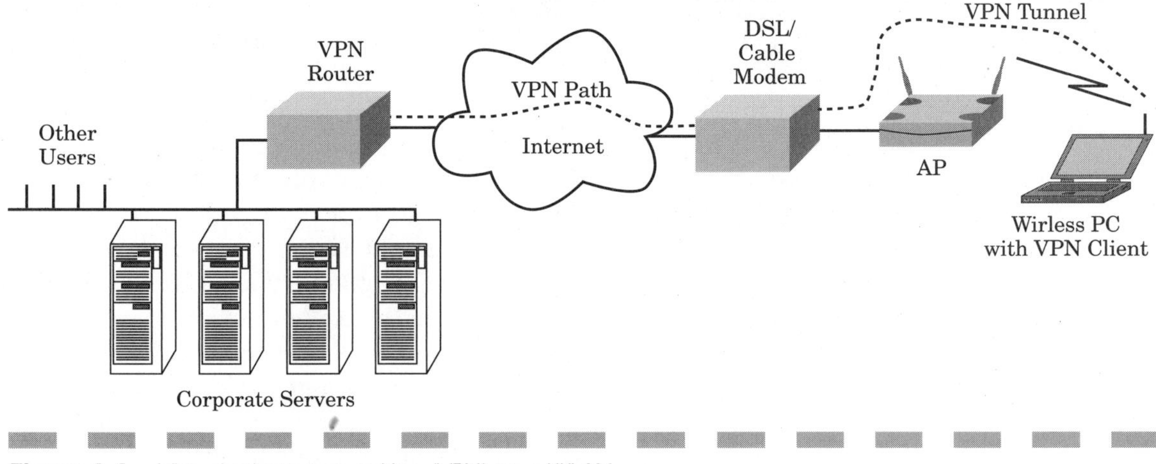

Figure 6.4 Virtual private networking (VPN) on a WLAN.

Figure 6.4 shows that the use of a VPN makes the problem of wireless security moot from the standpoint of intercepting transmissions to and from the AP. The only serious issue that remains is the authentication of a user's W-NIC before granting access to the AP. Listening to encoded wireless transmissions is about as useful as listening to a conversation in foreign language, if you are not knowledgeable of the language being spoken. The only difference is that absolutely nobody can translate the VPN transmission without the encryption key, and only the two VPN stations have that key.

In VPN terms, a secure *tunnel* is set up between the VPN end points. The concept of a tunnel is important in understanding the functionality of a VPN. A VPN's end points are the VPN controller or server at the corporate offices and the VPN client on your PC. The VPN server is often placed on an existing router or firewall. To initialize the tunnel, your client software requests a connection from the VPN server, which authenticates your request and sends you the appropriate keys to begin the secure dialog. Once this process is complete, your PC has a secure, virtual circuit across the Internet to your corporate office.

This VPN process does not need to be too sophisticated. A large company can use VPN technology to interconnect branch offices, home offices, and telecommuters. A small- to medium-size company can use a VPN just as easily for many of the same purposes. The value of the technology is that a secure connection is made through the virtual circuit. As a matter of fact, all forms of data can be sent through the same connection, including digitized voice.

A VPN can connect many virtual users to the central site. Depending on the particular model of VPN server you use, you could have 50 to 5000 tunnels. In general a tunnel is set up for each IP port over which you send data. An individual connection from your laptop could actually require multiple tunnels for your complete set of applications.

The current example shows only a WLAN supporting the VPN tunnel from one computer. Most installations will have at least several remote users connected by VPN. The only requirements are that the VPN server must be able to support the number of tunnels required and that the VPN server's Internet connection has sufficient bandwidth.

A type of VPN could be set up to encrypt your transmissions to the AP, but that is actually a simple data path encryption application. You can accomplish a wireless path encryption with the hardware encryption features of your W-NIC. Remember that the integrity of this type of secure connection depends on the frequency of key updates, the method of key exchange, the key length, and the strength of the encryption algorithm.

Firewalls

Another form of data security is the firewall (Figure 6.5). A data networking firewall is a device that ensures that only authorized connections are made into your network. The concept is rather like the bouncers that selectively allow patrons into a popular nightclub, except that your data doesn't have to be a trendsetter or a movie star. The firewall tries to ensure that only certain users inside the firewall can connect to certain locations outside the firewall on certain permitted IP ports.

Internet Protocol uses ports and sockets to identify certain types of communication and services. To talk about firewalls, we need to know a little about ports. Each port has a 2-byte address identifier. Thus, IP has been designed to recognize approximately 65,000 ports.[3] Many of the lower-numbered ports have been assigned familiar services. A list is shown in Table 6.3.

[3]The total number of possible ports is represented by 256 × 256 = 65,536. Ports 1–1024 are called the *well-known ports*, and those above 1024 are considered unknown ports. The well-known ports are sometimes called *server ports* and the unknown ports are called *client ports*.

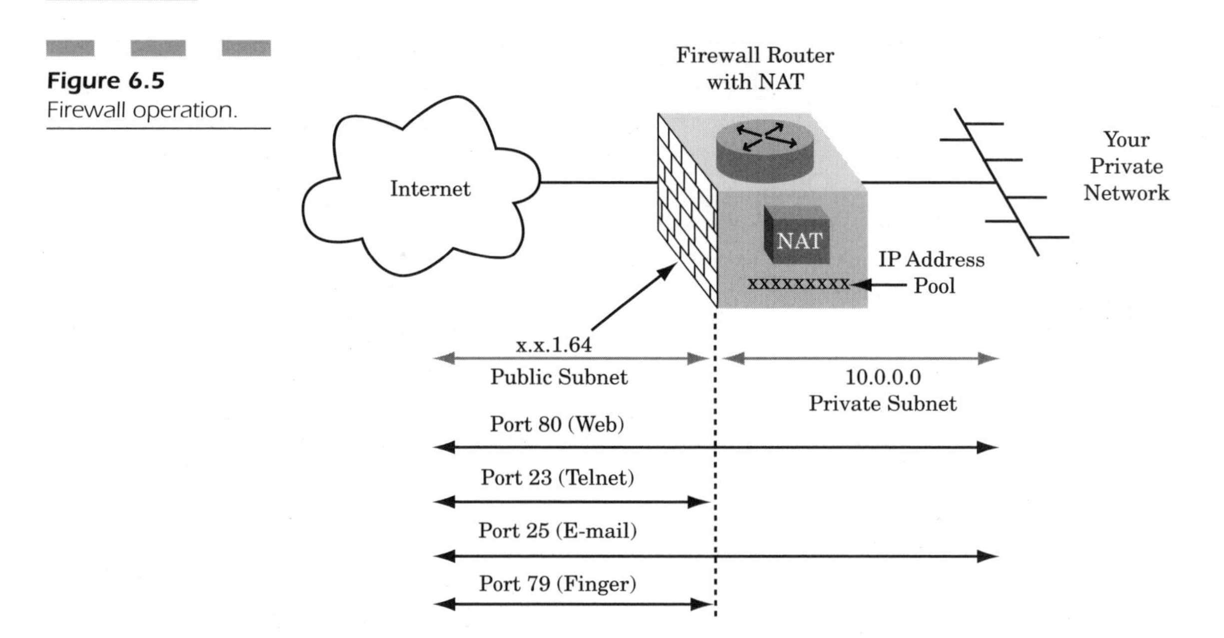

Figure 6.5 Firewall operation.

TABLE 6.3 Common IP Ports

Port Number	Service
7	Echo
21	FTP
23	TELNET
25	SMTP
53	DNS
79	Finger
80	HTTP
110	POP3
161	SNMP
162	SNMP Trap
446	SSL

In addition to the common ports shown in Table 6.3, many other ports are used by convention. The well-known port space was quickly filled up by various applications. Eventually, the port space above 1024 had to be used. Although the well-known ports were originally the only ports reserved for specific services, some standardization has been made for certain ports in the upper space. As new services have been created for the Internet, particular ports have been used by announcement and convention in the upper space. For example, audio and video streaming use certain upper ports, as do expanded network features for various OSs.

Many applications begin an IP conversation on the well-known port reserved for the corresponding use and then open up ports in the upper port space. This allows many IP sessions[4] to be supported from a single server over its IP address. As each session is begun, a predictable exchange occurs from server to client. Essentially, the client (your PC, for example) asks the server if it is OK to talk now and identifies itself. The server says, "Sure, what do you want?" The client says, "Give me the ____ web page," and the server says, "Here it is, up on port 25000," and sends the page.

A simple *static firewall* uses a packet inspection function that looks inside the IP packet to see what ports and IP addresses are being used. The firewall can disallow any communication that uses certain ports, such as Telnet, because of the security risk to the servers and computers inside the network. The management program for the firewall creates a filter to block these ports. Any packets that contain a disallowed port address are blocked (and discarded). The sending station never sees a reply. A static firewall also can block packets that contain particular IP addresses or address ranges. For example, the firewall could block packets to or from a known attacker's site. It also can prevent a device outside the firewall from pretending to be inside the firewall with the use of a fake, but valid, internal address.

Another type of firewall is called a *dynamic firewall*. Dynamic firewalls can selectively open ports that are being used by authorized services (and authorized hosts). A dynamic firewall monitors the initialization of a session and monitors the state of the session until it is closed. The firewall sees that the session is established and that it is using another port (or ports) in addition to the port that is primary for that

---

[4]You can view these port-to-port connection sessions by running netstat.exe from a command prompt in Windows. You will see the protocol, the local address (including port), the remote address also called the foreign address (including its port), and the state of the connection. This is also a way to observe surreptitious connections into your PC or server.

service. It opens the additional ports (allows packets through) for communication until the session terminates or until a timeout expires.

Firewalls also can use access control lists that consist of MAC addresses and/or IP addresses that are permitted to communicate through the firewall. This allows a network administrator to be very selective in granting access to internal resources.

NAT Firewalls

Firewalls (and some routers) may offer a feature called Network Address Translator (NAT). The firewall can substitute the IP addresses for your computers. (This is a little like the hero having an alias in a mystery novel.) NAT allows the users on the protected side of the firewall to disguise themselves to the outside world. The purpose is not to gain anonymity for its own sake but to prevent an intruder from gaining easy access inside a network by knowing (or by guessing) your IP addresses.

With NAT, a group of global IP addresses is placed in a pool by the firewall. When a computer on the inside of the network wants to reach the Internet, the firewall assigns an outside IP address from its pool. When each packet is sent across the firewall, the pool address is substituted for the original source address and then passed out to the Internet. The firewall keeps track of the corresponding internal address in a table. Any packet that is received as a response from servers on the Internet will have the substituted pool address as its destination address. The firewall will recognize the pool address and look up the corresponding internal address in its table. The packet is then modified to restore the real IP address and passed to the internal network, so it can reach the proper recipient PC inside the network. This process of address substitution is aptly called *network address translation*.

The NAT firewall acts as a router between the internal IP subnet and the external IP subnet, which in turn connects to the Internet backbone. If a subnet within one of the three private network addresses ranges (Table 6.4) is used, two advantages accrue. First, the use of these private addresses is unlimited. That means that you do not have to obtain the addresses from an Internet registrar, and the number of addresses is essentially unlimited for the purposes of a private network.[5] Second, the addresses are nonroutable. Routers reject the addresses, except for the type of network address translation to isolate our internal networks.

[5]Unlimited, that is, if 16 million or so is sufficient. The 10.0.0.0 class A address space alone comprises more than 16,700,000 addresses.

TABLE 6.4
Private Network Address Ranges

From	To
192.168.0.0	192.168.255.255
172.16.0.0	172.32.255.255
10.0.0.0	10.255.255.255

A fully functional NAT firewall can accept a very large pool of global IP addresses and provide a 1:1 translation to the internal private addresses. Alternatively, the firewall can do a many-to-one translation (or many-to-few). That mode of operation allows a reasonably large number of internal private addresses to be translated to a single external global IP address.

In fact, this many-to-one translation is exactly what is done by the simple NAT routers we often use for our SOHO networks.[6] It follows that this is a very rudimentary form of firewall operation. One of the major functions of a firewall is to prevent outsiders from accessing internal network resources, such as workstations and servers. NAT obscures the internal addresses entirely. They simply do not exist as far as the external Internet is concerned. Consequently, it is rather difficult to mount an attack to access those resources.

It is quite common for the router and NAT function to be contained within the same device in smaller SOHO networks. Some of these even support fairly simple firewalls. However, in large networks, it is more customary for the router to be separate from the firewall. In such cases, the firewall normally handles the NAT tasks, among other things.

For more information on private network address allocations and the operation of NAT, consult the Appendix.

Personal Firewalls

Over the past several years, it has become painfully obvious that border firewalls, as covered in the last section, are not a cure-all for network intrusion. Very serious problems have been experienced as a result of computer viruses and unwelcome *microservers* or *responders*. Most of us are familiar with computer viruses, but the concept of a microserver is a

[6]See Chapter 4 for more information on SOHO NAT routers.

new one. This is a term for a hidden program that is placed on a PC, intentionally or surreptitiously, that responds to directed inquiries. The microserver may deliver information to a remote destination device or give a remote user control over portions of the unsuspecting user's PC.

These routines are increasingly clever and innovative, as their perpetrators are scurrilous. The purpose is sometimes innocent: an application checking with its home site for updates or anonymously reporting statistics. However, these tiny programs often are used to launch mass attacks on popular Web servers or to capture and report keystrokes to a malevolent party. One type of program is called a Trojan Horse, after the ancient technique to invade from within. Such a program plants an invisible process on your computer to perform some malicious task.

Frequently, a microserver or Trojan on one of the computers on our internal network will turn against its own. Such a "plant" will start attacking the other computers on its subnet, looking for a point of entry. Once found, the little devil will clone itself to the victim computer. Then the process begins again.

In the class of benign hidden routines are the "phone home" automatons. A software vendor determines that it would be of benefit to frequently inquire whether there are updates or new versions of which you should be aware. This is a truly self-serving behavior because the vendor hopes and expects that you will download an upgrade to your existing product. This produces more revenue for the application vendor and thus is the raison d' etre.

A new class of firewall program, called the *personal firewall*, has become available. This firewall product is a software application that screens all incoming and outgoing Ethernet traffic on our individual computers, as illustrated in Figure 6.6. Several personal firewall programs are available, with a range of capabilities. Most are easy to set up, allowing the user to choose their desired level of security. In general, the programs are well behaved and effective.

Personal firewalls are relatively inexpensive to purchase. Some personal firewalls come as part of a virus protection suite or as part of the OS. Windows XP and successors, for example, have built-in firewalls that may be enabled for the protection of the computer on which they are installed. But be cautious: a firewall offered by an OS vendor may not be the best approach.

It is now an accepted fact that we must protect our PCs against attacks from outside and inside our own network connections. The only practical way to accomplish this is to install a personal firewall on our computers. Fortunately, a number of software manufacturers have answered the call.

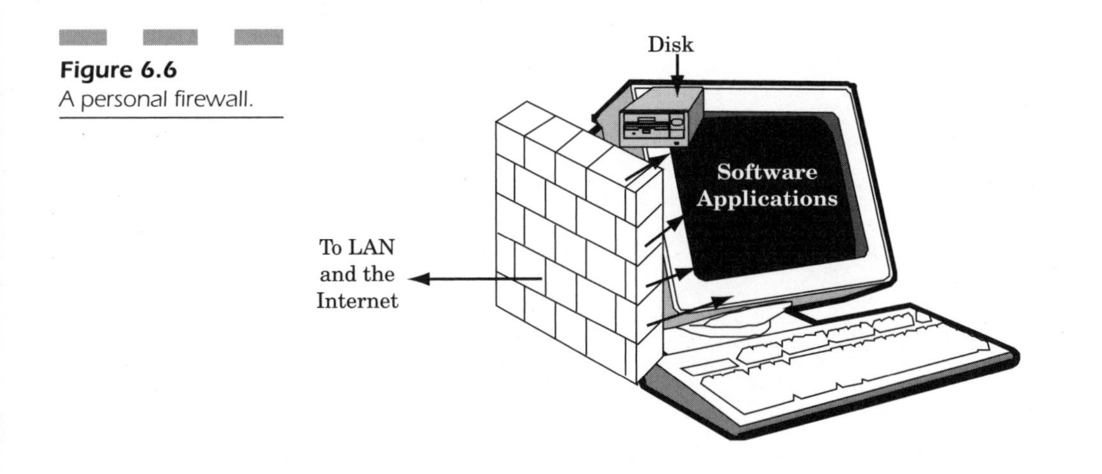

Figure 6.6
A personal firewall.

A variety of personal firewall products are suitable for installation on Windows, Mac, and Linux systems. The greatest variety is available for Windows, probably because those systems suffer the most. It would be impractical to list all of the personal firewall products here, because the list would soon be out of date. The field of players is growing rapidly. For a current list, consult www.BuildYourOwnWirelessNetwork.com.

The High-Tech Home

Chapter Highlights

- WLANs in the High Tech Home
- Residential Wiring Standards
- Phone, Data, TV, Security
- Entertainment
- Home Automation
- Bluetooth and HomeRF
- CEBus and HomePlug

Modern technology has begun to make tremendous changes in the way we design and use our homes. In many ways, this revolution began early in the last century, if not in the latter decades of the century before that. This is not to say that the technology advances in the past two decades have not been amazing. Clearly, the very rapid onset of new technology has been dramatic, but the introduction of technology has happened before, just at a slower pace.

To gain some perspective, let's mention a few of the changes in home life that have occurred as a result of the innovations of the Industrial Age. If you pictured a home of about 150 year ago, you would notice that many things would be missing that we take for granted. For example, there would be no electric lights, no telephone, certainly no television or radio, no central heat, or air conditioning. Good grief, there would be no garage door openers...what would be the use, because it would just scare the horses?

Fast-forward about 40 to 50 years, and you would be amazed to be able to press a switch and have a light spring to life, through the magic of electricity. And you could actually speak to another person some distance away through a wonderful invention called the telephone. As you would expect, more and more people began to demand the convenience of these new inventions. Rural electrification and telephone co-ops sprung up to provide these new-fangled conveniences to homes away from the cities.

Time travel forward another 50 years, and you would take electric wiring and telephones for granted. Hadn't they always been around? It certainly would seem so for the young. Older houses had to be retrofitted with electrical wiring and telephone cables to make this happen, but new homes were built with all the modern conveniences. Television was new, and it was broadcast through the airwaves, as was radio. Stereo was just emerging, along with FM radio. What a brave new world!

Over the distance between the 1950s and now, some radical changes have happened, technologically speaking. The most significant single force has certainly been the incredible miniaturization that has resulted from the invention of the transistor and, later, the integrated circuit and the microprocessor. Developments in microprocessors, digital transmission, telecommunications, and remote control have given us a host of new devices. Many of these devices have found a place of importance in our homes.

Now that we have so many of these technical innovations available to us, we have rediscovered an age-old problem ... at least for the new age. Our homes need some technology updates to allow these devices to function optimally. However, instead of rewiring for electricity, gas, or phones, we are rewiring for computers, cable TV, and security. Again, new homes are designed to incorporate many of these new features, as we have now come to depend on them for our High Tech Home.

Supporting WLANS, Telephone/DSL Service, and Cable TV in the High Tech Home

It goes without saying, in a book on WLANs, that the first technological innovation you must consider is a wireless computer connection. We have spent much of this book describing the technology to support a SOHO WLAN. It is clear that your home should include this feature to be completely high tech.

Unlike many of the other technologies that are covered in this chapter, you do not need a lot of wiring and cabling to support the WLAN, but you do need some. For one thing, to install a functioning WLAN, you need a connection to the Internet. This connection must be wired to a DSL or a cable modem, and that modem in turn must be connected to the appropriate service provider through a telephone line or a cable television line.

Both high-speed connections, cable and DSL, require excellent connectivity to operate properly. When you first have DSL or cable modem service installed, you probably will be expected to provide a new connection between the modem and the service entrance. The term *service entrance* is the point of entry to your residence. For telephone service, this is often called the *point of demarcation*, or the *demarc*,[1] for short. For cable service, the service entrance is typically where the lightning arrester is mounted and the entrance cable is terminated.

Telephone Service

The telephone service entrance is where the telephone cable from the utility pole or the utility junction box terminates at your house. If you have a service cable strung to your house through the air (appropriately

[1]This is sometimes spelled demark. From a legalistic standpoint, the point of demarcation is where the service provider's legal responsibility ends and the subscriber's responsibility begins. In the United States, this change became significant after deregulation in the 1980s. You may have to pay for installation and troubleshooting service to wire past the point of demarcation.

called an *aerial cable*), there will be a strain-relief mounting on or under the eave, usually at a side wall, and a junction box where the telephone wires are terminated. If you have underground service, the entrance cable will come up from the ground to the junction box.

Over the past several years, most phone companies have replaced the simple four-wire terminal box with a more sophisticated termination called a Standard Network Interface (SNI).[2] The SNI is generally a gray plastic box (Figure 7.1) containing the incoming cable terminations, the surge protection, and the disconnect plug. This box allows the owner to connect interior phone cabling (premise wiring) to the box without disturbing the telephone network connection. It also allows the premise wire to be temporarily disconnected from the network for testing. In the age of deregulation, the phone company's responsibility ends at the demarc (unless you pay for interior wire coverage), and you are responsible for all of the interior cables from that point on.

Figure 7.1
Telephone Standard Network Interface (SNI).

[2]The Standard Network Interface also goes by a number of other similar names, such as Network Interface Demarc (NID) or "that gray box."

If you live in a multi-unit building, your phone service entrance will be in a central location for the entire building. The telephone company's point of demarcation may be at the service entrance or in a telephone interconnection closet on each floor. In many cases, it is your responsibility (or the building management's) to provide all the wiring back to your apartment or flat. However, you normally will have a single location in your unit from which the phone cables are distributed. Commercial office buildings are wired and administered in a similar manner. However, condominiums, which are separately owned, may each have a demarc on the exterior or the interior of the building unit.

To provide the very best DSL service, you will need to provide a new (or clean) direct line from the junction box to the DSL modem. A DSL filter is connected at or near the junction box to separate the voice frequencies from the DSL frequencies, so neither interferes (Figure 7.2). This method is called a *centralized filter DSL line*. Remember that DSL service is provided on an ordinary (usually existing) phone line. If you provide centralized wiring in your home, as described in a later section, you can place this filter in your telephone distribution panel.

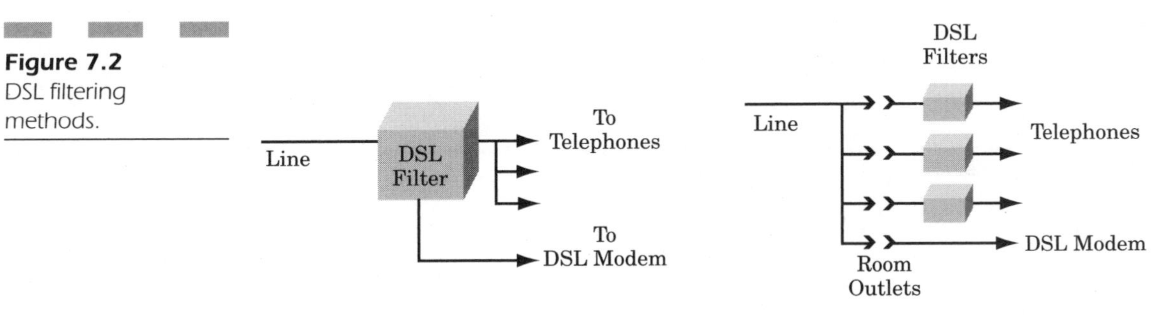

Figure 7.2 DSL filtering methods.

DSL service also may be provided without the centralized filter. This method is preferred for the "self-installed" kits that DSL providers are fond of promoting. With this distributed filter DSL installation method, you have a DSL block filter for each telephone. The blocking filter is a short section of phone cable with an integral DSL filter (Figure 7.3). If it is difficult to run a clean line for your DSL connection, this is a potential alternative. However, the distributed filter method can create poor DSL performance in a residence with many telephones and interconnecting (daisy-chained) lines. The DSL signal tends to wander up all these branched lines and get bounced around. The reflected signals create

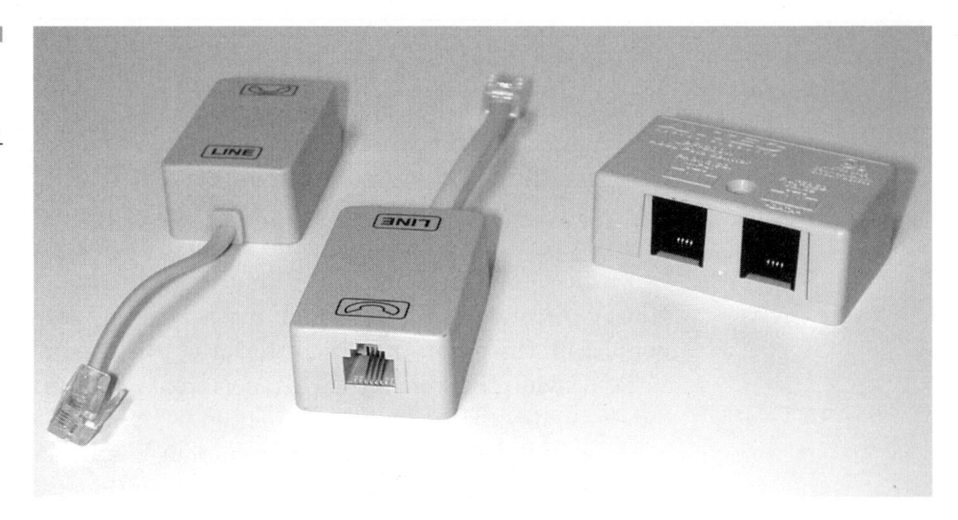

Figure 7.3 Distributed and centralized DSL filters.

interference in the same way that an antenna that is not adjusted properly creates annoying shadow on broadcast TV.

The best practice, if it is at all possible, is to isolate the cable for the DSL modem from the other telephone cables in the house. The most practical way to do this is to use the central filter method and provide a new cable just for the DSL service. As an alternative, you could use one existing phone cable to connect the DSL modem and a single phone. To do this, you would filter block all the other cables at the SNI and provide a distributed filter near the outlet where you intend to connect the DSL modem. Connect a two-line adapter at the outlet and connect your telephone through one of the distributed filters. Connect the DSL modem, unfiltered, to the two-line adapter.

To prevent the inadvertent mixing of filtered and unfiltered connections, some DSL modem manufacturers use the Line 2 position on the RJ-11–style jack for the DSL line. This is illustrated in Figure 7.4. A standard analog telephone line normally is connected on Pins 3 and 4 (called Pair 1) of the RJ-11 jack/plug interface. It has been this way for all time. However, data equipment manufacturers (and the "phone company") often used other pin combinations and even other jacks to keep data and telephone signals (and voltages) apart. Some DSL manufacturers wire their DSL modems so that the active pins are 2 and 5. In telephone jargon, this is Pair 2.

If you have a special jack wired for DSL only, it will use Pair 2. A standard, four-wire, telephone cord has the proper pins connected to support phone or DSL connections, and thus can be used to connect the DSL-only jack to the DSL modem. The other wires in the cable and their

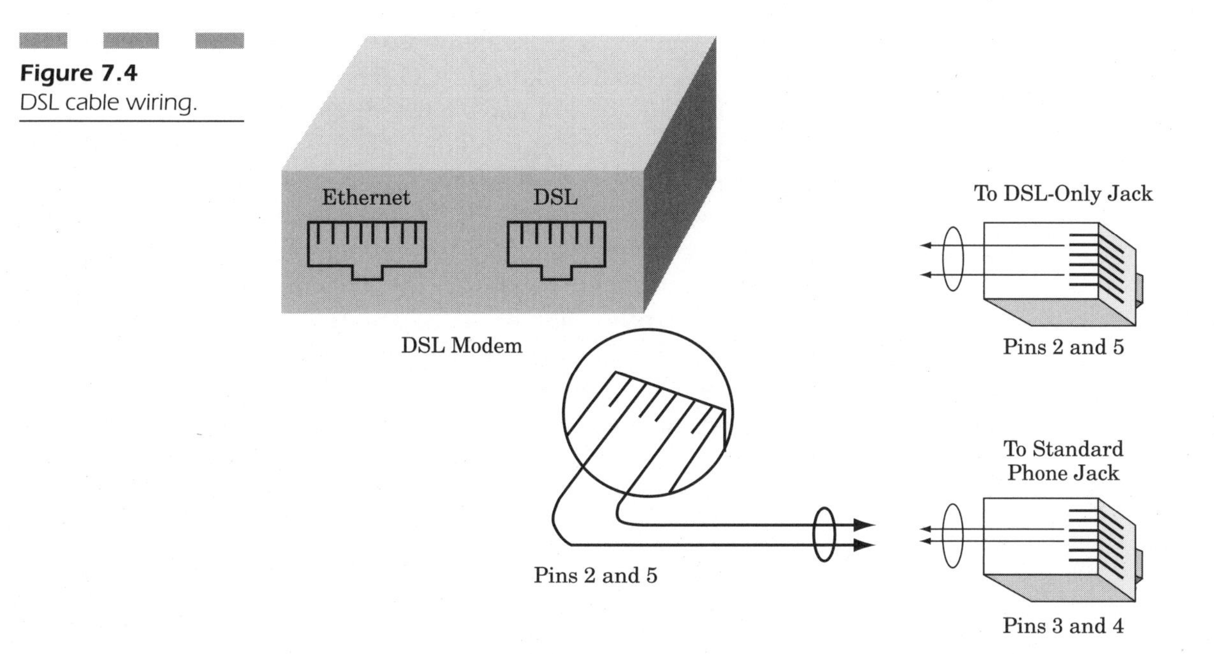

Figure 7.4
DSL cable wiring.

respective pins are ignored. However, if you try to connect one of these DSL modems to an ordinary phone jack, the connection will fail unless you use a special adapter cable that converts Pair 1 to Pair 2 (Pins 3–4 to Pins 2–5) This is not a plot to make you crazy.[3] This helps assure that the proper types of cable are connected to the DSL modem and line and that filters are used to isolate the DSL and analog phone signals. Nevertheless, if you are not aware of the subtle differences in these connections, you can run into trouble. Fortunately, the proper adapter cable is usually a distinctive color, which helps a lot.

Inside Wire

Another point should be brought up here regarding conventional home telephone wiring. In the days of single-line analog-only service, many homes were wired using straight (unpaired) multiwire cable. The various local exchange carriers were once part of a small number of monoliths, who realized that this untwisted wire was slightly cheaper than twisted pair cable. The major reason for twisting the two wires of each

[3]Yes it is. *No it isn't*. Yes it is! *No it isn't*! Well, it is driving us crazy!!!

pair together, of course, is that many user lines can be combined into larger cables to be carried back to the CO. The twisting minimizes cross-talk from pair to pair, so you don't hear all those other conversations. (One or two might be interesting, but a cable full would be unintelligible and interfere with your own conversation.)

However, there was no need to isolate phone lines in the single-line homes that were common years ago, so the simpler, untwisted, type of wire was used. This cable is generally called *inside wire* (IW), but it also goes by a number of other names, including JK (for the terminal markings in the early telephones) and four-wire (not two-pair, because the wires are not twisted into pairs). It consists of four individual wires, normally colored (and named) green, red, black, and yellow. The green and the red wires are used for the analog phone signal. Originally, the black and yellow wires provided ground and ring voltage but later were used to power the little lamps in early tone telephones. (The lighting is now provided by a line transformer or by the line voltage itself.) However, when the installation of a second phone line became popular, it seemed natural to use the other two wires to connect the second pair. Oh, boy, did that cause problems! Suddenly, there was cross-talk so severe that you could virtually talk to the party on both lines. Teenagers went berserk, at least when they realized you were listening.

The cure turned out to be to use twisted pair cabling as inside wire for any homes that had (or expected to have) two or more lines. This eliminated virtually all cross-talk between phone lines.[4] In addition to the voice-frequency cross-talk problem, conventional nontwisted IW created havoc for the high frequencies of a DSL signal. This is why the knee-jerk recommendation for DSL lines is to run a separate, new, clean, *twisted-pair* line back to the demarc. The twisted pair cable avoids much of the signal loss and cross-talk problems. The DSL provider really has no idea what sort of inside wire you have, so this advice plays it safe.

Cable TV Service

In many ways, cable modem service is much simpler. Only coaxial cable is used to distribute the cable signals and there is only one way to wire

[4]Cross-talk can still be created by flat (silver satin) phone cords that have more than two wires. What happens is that signal coupling occurs between the adjacent wires in the cable, so if the jack is wired for both lines, you can get cross-talk in a four- or six-wire flat cable, even though the phone (or answering machine) connects only to one pair.

the connector. However, there are several rules of operation that you must follow to successfully install a cable modem in your home.

The first thing to understand about cable modems is that they use vacant cable channel bandwidth to transmit signals up and down the line. That is, the cable provider sends a data signal to you (and others on your cable segment) on one set of frequencies, and you send data back on another set of frequencies. On most systems, these data paths occupy channel space at the opposite ends of the cable spectrum. We won't go into too much detail here because cable modem operation was covered in Chapter 3. However, from the standpoint of home wiring, the use of the upper channels places a real performance strain on the coaxial cable that connects your cable modem.

One of the ways in which cable (or CATV) signals are distributed in a home or apartment is to *split* the signal for each outlet (and thus each TV). A standard coaxial signal splitter has one input and from two to four outputs (Figure 7.5, top). In this example, an optional CATV signal amplifier is shown, with a four-way splitter and a downline two-way splitter. At each splitter, the received signal is reduced by the splitting plus an insertion loss. Notice that the cable modem is connected to a clean line and a new two-way splitter added at the service entrance. This avoids the problems posed by the splitter losses, cable losses, and the amplifier. Unless you have one of the new bidirectional amplifiers, it is doubtful that the amplifier would allow any reverse signal from the cable modem. So, even if you could get enough receive signal, your cable modem could listen, but it couldn't talk.

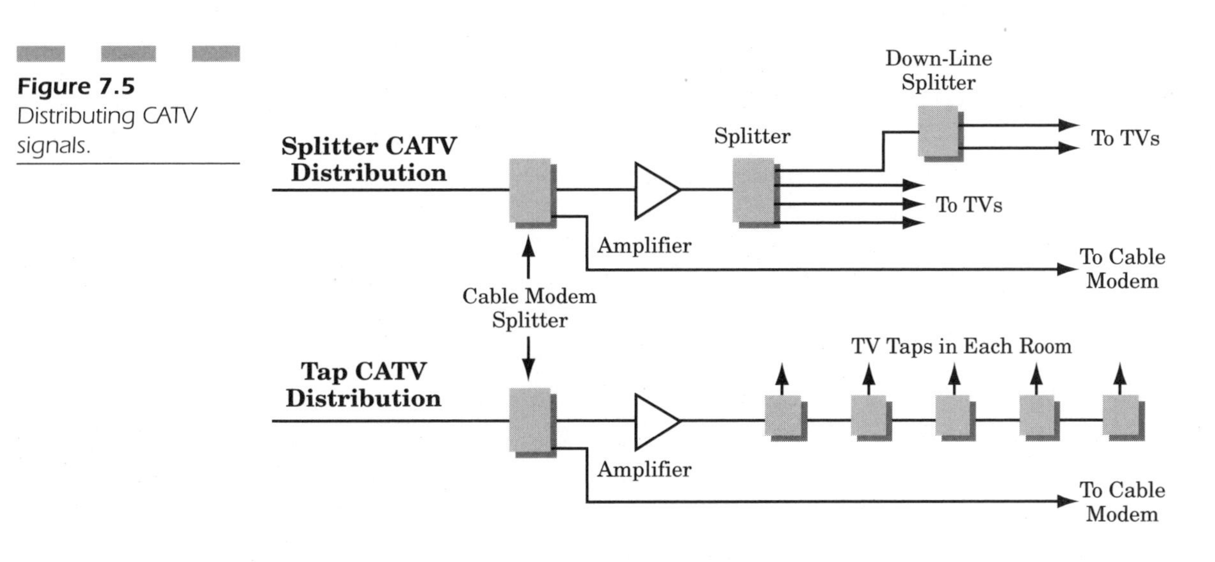

Figure 7.5 Distributing CATV signals.

Another way to distribute CATV signals is to use a tapped line. Actually, the line is not tapped. The cable is simply run from outlet to outlet in a daisy-chain fashion. At each outlet, the incoming cable is connected to a type of two-way splitter called a *tap*. The tap takes a tiny portion of the signal[5] and passes it to the tap outlet. The balance of the signal is passed to a "pass-through" connector, where another coax cable runs on to the next outlet. At the last outlet, a terminator is used (if it is done right) to eliminate reflections that cause image ghosting. There are several problems with this technique. The first is that, if a cable modem is connected to one of the taps, it will suffer a considerable signal loss in both directions. In addition, the unidirectional amplifier, if used, would block the return path, so the modem could not connect back to the provider. Placing a splitter at the service entrance, before any of the amplifiers or taps (as shown in Figure 7.5), allows the cable modem to connect with a robust, two-way signal.

Residential Cabling Standards

The advent of telephone, data, and television signal distribution in the high tech home has given rise to a number of global and proprietary wiring standards and practices. Believe it or not, there are a number of approved national and international standards for residential cabling. Among these is the *ANSI/TIA/EIA 570-A, Residential Telecommunications Cabling Standard*. This standard provides recommendations for single and multifamily residences. Included are all sorts of applications for the high tech home, from data and voice to entertainment and security. Many technologies for features such as temperature control and intercom can be considered primarily as convenience features. But then, are not all aspects of high tech homes mostly for our convenience and enjoyment?

The most common conceptual element in all public and proprietary standards for residential cabling is the concept of a central distribution system (Figure 7.6). Basically, this scheme provides a central location where all of the outside services are brought into the house, including

[5]Taps are usually set for a –20 dB split to the tap connector. System signal levels are set to provide adequate signal to the last outlet, allowing for an almost negligible tap and connection loss at each tap. This technique also saves cable compared with the home-run method of centralized cabling but eliminates any flexibility for TV signal distribution to individual devices.

telephone lines, cable TV, and data lines. From this central point, cables are run for each service to each planned service location (or outlet), without splices or taps. These cables are normally installed in a direct, one-for-one fashion, called a *home run*. (If you don't know what a home run is, two examples are when you hit the baseball out of the park, and when you turn your horse toward the stable, where the hay is. Both get you back home pretty quickly.)

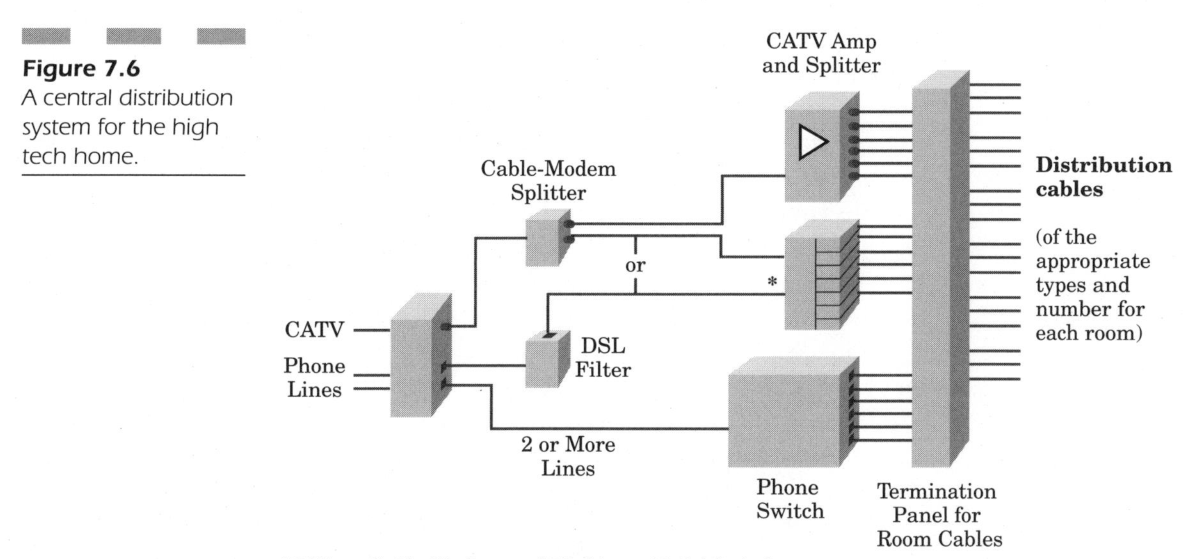

Figure 7.6 A central distribution system for the high tech home.

The central wiring location is also the logical location to place common premise distribution equipment, such as the CATV amplifier/splitter, the DSL/cable modem and filter, a router and multiport hub, the security system, and the multiline telephone switch, if one is used. In a multifamily residence, a form of structured cabling is used that is similar to what you would install in a commercial office building. In a multifamily building, the services are delivered to a central point, but multiple distribution cables branch out from there to intermediate distribution points. These points are smaller distribution locations in each building or floor in a residential complex. From that point, the signals are distributed to each residential unit, which ideally has its own internal centralized distribution system.

Certain services to the multifamily residence, such as telephone and data, are cross-connected in each intermediate location. Other services, such as CATV, are usually amplified and split according to the level of service for which the subscriber has contracted. CATV distribution equipment must provide for two-way transmission to allow cable modem operation. Cable modem services and DSL services may be provided by a third party who installs special equipment for the service to be provided.

Wiring the Home

For the home of 20 years ago, there were only two services that needed special cabling: telephones and cable TV. In many areas, cable TV was not available or was not widely installed. However, in today's world, these two services are almost *de rigueur*—a required feature. And the same will soon be said of two more services, data and security.

The thoroughly modern, high tech home needs all four of these primary services, and we would add a fifth. We call these communication services—telephone, TV, data, and security—the Big Four. The fifth dimension (or element, if you like) is entertainment.

This section briefly describes the considerations for providing all these services in the high tech home. We can see that a centralized distribution network is crucial to effectively deploying these services throughout the home. A comprehensive central wiring system is the best way to ensure that telephone service, computer connectivity, and the other services are available wherever they are needed. It is somewhat analogous to the central nervous system. The home distribution center is the brain of our high tech system. It is where all of the interconnections are made and from which all the outputs of the system are sent.

Telephone

The basic cabling concepts for fixed wiring for telephone and data are very similar. They differ only in degree. Telephone and data both use twisted-pair cables, with two to four pairs of wires in each cable. However, from that point on, everything is different.

Telephone devices generally use a six-wire RJ-11 connector. This connector can support up to three pairs (two wires each) and is the standard for the single line phone. One pair (two wires) is all that is required for an analog phone line, but almost all modern cabling uses at least three pairs. In the past, telephone cabling was run to the nearest outlet location and spliced to go on to the next outlet. This continued in one or more additional runs, in a daisy-chain fashion, until all of the desired phone outlets were wired. The inside of the house was like one giant party line, with all the outlets hooked together. If a second line was needed, another pair could be tapped, and if the outlet was wired properly, a two-line telephone could be connected.

Over the past several years, the popularity of home offices and second lines for children or a parent has made this old standard telephone wiring somewhat obsolete. The new standard for telephone wiring uses home-run cables from each phone outlet to a central panel. It is now possible (and quite affordable) to connect individual telephone instruments to a miniature phone switch, similar to the larger PBX switches that businesses use. Several different types of telephone switches are available for SOHO use. These may allow four or more CO lines (commonly called phone lines by the public) to be connected to as many as eight or more telephones (properly called *stations* or *instruments*). Some systems can accommodate eight incoming lines and 32 stations. If your house is not that big, don't feel alone.

Some of the conveniences these systems offer include: hold, call transfer, intercom, and centralized voice mail. The prices on small phone switches is dropping dramatically, and their use is greatly facilitated by a centralized phone cable distribution system. The distribution center for your cabling is the logical place for your phone switch.

Telephone cabling should use a minimum of TIA Category 3 cable.[6] Category 3 uses 4-pair cable that is tested to meet certain performance standards. Telephone systems also can be supported on four-pair Category 5e cable. Category 3 and 5e standards specify eight-pin connectors, but six-pin modular connectors are available for ordinary telephone use.

Telephone, data, and closed circuit TV (CCTV) services may be connected at the same outlet box by using modular connector inserts. Figure 7.7 shows an example of a modular communications outlet plate with several connector options. The standards conservatively call for two cables to each outlet, for telephone and data. However, in the high tech home, you will want to provide at least three and possibly four cables. These would be used as follows: one for telephone, one for data, one for TV, and one for a second phone line or fax line.

Locations where you expect to have multiple devices may call for more cabling. For example, if you expect to have a network server or an entertainment center (with TV, DVD, VCR, etc.), you could provide additional cables of the appropriate type to that location.

---

[6]See *LAN Wiring*, 2nd ed., for more information on Category 3 and Category 5e cabling and the cabling standards.

Figure 7.7
A multiple-use communications outlet suitable for the high tech home.

Data

Data cabling layout is fairly similar to modern home-run telephone cabling. Data cables (station cables) are run directly from each outlet to the distribution center. At the distribution center, each station cable is terminated, often in a patch panel. Patch cords interconnect the station cables to the appropriate data service.

At the distribution center, you can install an Ethernet hub (or switch), a DSL or cable modem, and the DSL filter or a splitter for the cable modem line. If you use a router (which we recommend), this is a good location for it. Some modems are combination units with a router and a four- or six-port hub built in. You may want to use a separate data hub or leave space to add one later. Data applications are proliferating, and, by some predictions, all our appliances will soon need data network connections. Can you imagine it? Your refrigerator can connect to the Internet grocer and order some milk. But can you really trust it to know your tastes and brand preferences, or will it have a mind of its own?

The standards for data cabling call for a minimum of Category 5e in most cases.[7] Fiber optic cabling is also part of the standard. In addition,

[7]Category 5e, the extended performance specification, has replaced ordinary Category 5. The enhanced capabilities allow the cable to support Gigabit Ethernet over copper.

the standards for Category 6 and 7 components are on the horizon, and some of the components are available now. It is becoming difficult to keep up with all these new enhancements, but we do want our data network cabling to keep up with the times.

One way to keep up with the changing cabling standards is to provide your home runs within runs of plastic tubing. If the tubing runs are continuous and the turns and bends are not too arduous, you can change the cables at a later (and we hope distant) date. This allows you to support almost any future communications wiring enhancement.

It would not be proper, in this book on WLANs, to not mention the role of wireless networking in the high tech home. A home WLAN is the easiest way to provide high-speed connectivity anywhere in the home. The beauty of the WLAN approach is that it can be applied just as easily to older homes as to new ones. Even if you wire to every room in the home, or even two walls in every room in the home, you cannot sit at the kitchen table or in your easy chair without a data cable strung to the wall. The WLAN lets you connect wirelessly with ease.

Use the prewired data outlets to support your fixed computing assets, such as desktop computers and servers. Oh, yes, and that surfing refrigerator, if you have one. Also, you can use the data outlets to connect your wireless APs, so you will have wireless everywhere. Now, that's the life!

TV and CCTV

Television signals may be provided through a common antenna, through a cable television provider, or by a satellite system. Let's look at each of these television sources individually.

Reception of broadcast stations requires an outside antenna and probably a distribution amplifier and signal splitter to get the signals to each TV. Outside antennas may be equipped for coaxial cable that is routed into your distribution center. From that point on, the television signal is treated the same as any cable TV signal. Simply connect the splitter input to the antenna cable and optional amplifier.

If you use an outside antenna, you should pay particular attention to lightning protection. The other services, telephone, CATV, and data will normally have their own lightning and surge protection. However, you will have to provide your own lightning arrestor, drain wire, and grounding rod for your outside antenna. As an alternative, you might consider placing the antenna inside your attic. This is not an absolute cure for the surge problem, but it is a good option and it can meet the requirements of restrictive covenants against outside antennas.

Regular cable television is often called CATV, which is an acronym for community antenna television. Nowadays, "communities" are apparently pretty large, and some cable TV systems span entire metropolitan areas. A CATV provider brings television signals to a house on a coaxial cable.[8] With a little friendly persuasion (it's not all that difficult), the installer can terminate the cable in your distribution center. From there you can connect it to a splitter and optional distribution amplifier and a two-way splitter for your cable modem.

Even in this close environment, it is best to maximize the cable modem's signal by keeping it off the main splitter. Plus, many distribution amplifiers are incompatible with two-way signals. By the way, this may be a consideration if you plan to subscribe to a CATV service that requires two-way signaling to get cable-on-demand or other services. Two-way amplifiers are available, but the cost is somewhat higher than for one-way units, and the availability is limited. Even if you are not using two-way CATV services now, you should install two-way CATV amplifiers in a new centralized distribution system.

A satellite dish is the third way you might get your TV signals. With this type of system, you mount a dish antenna outside and run the cable from the dish to your distribution center. A satellite system adds some complexity to signal distribution because the satellite signal is down converted at the dish, and an intermediate signal is sent to one or more receivers. These satellite receivers (frequently called a *tuner* or *satellite box*) then convert the tuned digital channel to an analog TV channel so you can watch it on your television. This system is not necessarily compatible with amplifiers and splitters designed for ordinary CATV. If you plan to distribute satellite TV signals over your centralized distribution system, be sure to use cable, amplifiers, and splitters designed for satellite TV signals.

Security

Security cabling is another important item for centralized cabling in the high tech home. In recent years, home security systems have become a very desirable feature. Some would say a home security system is almost necessary. Certainly, home security systems increase the home-

[8]Some CATV providers are now using fiber optics for signal distribution.

owner's protection from break-ins and can be interconnected to smoke detectors and fire alarms. Home security systems can also be equipped with video cameras for added security. In addition, a home security system may be provided with remote listening and remote camera devices that can be accessed by the alarm company.

In many new high tech homes, security systems are prewired to a security panel. The security panel is often in the master bedroom or a central utility closet. This is a logical location for the central distribution panel for your other high-tech features, such as telephone, data, and cable TV. As a matter of fact, some distribution panels provide interconnection of security video cameras. Most security installers, however, will provide a separate enclosure for mounting the security controller and the terminations for security wiring.

With the tremendous amount of wiring for perimeter alarm contacts, security cabling is a natural for prewiring to your central distribution point. Security cabling typically involves a combination of home-run and daisy-chained wiring. Some of the security connections are naturals for daisy-chain wiring. For example, to connect a series of perimeter contacts for outside windows, it would not make much sense, economically, to use home run cabling from each window. However, connections to door contacts are normally made with home run cables, because door openings and closings are important in the operation of a perimeter security system. Often the security system does not arm itself until there is a door closure or specified delay. Entry delay is also tied to specific doors, whereas other door openings may cause an immediate alarm. Glass-break detectors also generate immediate alarms and frequently are run separately to the alarm panel.

The high tech home also may use CCTV cameras. These cameras require coaxial cable that is very similar to that required for CATV. A typical high tech home can have four or more CCTV cameras. One camera might be located at the front door, another at the back door, one covering the driveway, and another covering the pool. Some parents may want a camera in a baby's room or a utility area.

Most security cameras are line-powered, which means they get voltage over the coax cable that also sends signals back the video signal. Security cameras usually connect to a central control unit in the home. Some camera systems may interconnect to your data network, so you can see the image on your computer's monitor. It is also possible for the alarm company to provide a video or audio connection into your home. This allows the alarm company to view the video or listen to the audio remotely and eliminate innocent causes from resulting in a false alarm.

This is a very valuable feature for commercial security systems, but many homeowners would consider it a potential invasion of privacy.

Self-contained video camera systems are also available with a data connection. These units are very compact, but do require data and power connections. This type of networked camera is often called a *Web-cam* because it has a built-in HTTP server. You simply point your Web browser to the IP address of the Web-cam and view the image. Unlike CCTV cameras, these Web-cams take pictures at fixed intervals, such as once per second, to prevent loading the network with image packets. If you are connecting remotely to a Web-cam, it is important to reduce the amount of data being transmitted by the camera. Web-cams typically capture images in a *jpeg* format, which is native to Web applications. A jpeg file with normal resolution is about 400 to 500 kB, which can take quite a while to send over a dial up connection. However, if you connect over a LAN, you may be able to tolerate a much higher image rate. Most Web-cams can capture images at up to 15 times a second. This rate gives you almost continuous motion, although it is still jerkier than a normal CCTV camera image.

Entertainment—The Fifth Dimension[9]

We have covered the four conventional home technology services: telephone, television, data networking, and security. In the past few years, a fifth technology, entertainment, has emerged. Home entertainment systems have moved far beyond the simple stereo system. The two hot technologies in home entertainment are audio distribution and home theater systems. No truly high tech home would be complete without these systems, in addition to the four basic ones.

The most important thing to realize about home entertainment systems is that there are two distinct ways of installing them. One way is to follow the centralized distribution system that we have just covered. To wire your audio and video distribution in this manner, you would make a home run of the appropriate cables back to the central distribution location, where you would make the required interconnections to your audio and video systems. Unfortunately, this may not be the most effective way to wire your system.

Audio and video entertainment systems normally have a number of bulky components at the main (or central) location. A typical audio sys-

[9]With apologies to the 5th Dimension, the hit vocal group of the 1960s and 1970s.

tem will have a variety of components, such as a tuner, a graphic equalizer, a CD changer, a tape deck, a receiver, a pre-amp, and a power amplifier. One or more of these components may be combined, but you still have lots of interconnecting cables and cabinetry. Further, you usually need to have all this audio equipment in a highly accessible location, often where you normally enjoy using the system. If you connect to other audio systems in other parts of the house or distribute the outputs to speakers in other rooms, you will need a speaker switch and volume controls at the central location, and perhaps in each covered room. None of these items is particularly well suited for mounting in or near the central distribution panel that you use for other high tech services.

Video equipment, including home theater equipment, has the same problem. Mostly, you need to have localized, in-the-wall, wiring that connects the components of your home theater. This would include projection TV, surround-sound speakers, and lighting controls. You also have all the video components, including the cable/satellite converter, a VCR, a DVD player, and various video accessories. In most instances, the audio system and the home theater system are interconnected. Often, these video components need connection back to the centralized CATV distribution system, so you can watch a movie in another room than the VCR or DVD player.

So, what is really needed is a secondary cabling system to support home entertainment. This system typically concentrates on the primary media room, which is often a family room, den, or even a special entertainment room. In this entertainment room, you will need speaker and projection video cabling, connections between the audio and video system, and one or more connections to the central distribution panel. You also will need enclosures for switching, input configuration, amplifiers, and special adapters.

The basic interface from a CATV or satellite system to your video equipment is over a standard TV channel, often channel 3 or 4. However, some systems may provide digital TV interfaces. Several newer video interconnection standards have emerged, including S-Video and fiber optic interfaces. In addition, almost all systems can connect using separate RGB[10] and audio cables or using baseband NTSC video. Most video systems interconnect their components using the RGB interface and discrete audio connections for right-channel, left-channel, and (if used) center-channel. Specially prepared cables must be used for these audio and

---

[10]RGB video uses three video cables, one each for the red, green, and blue video channels. The signal is baseband, that is, unmodulated by an RF carrier, and it is at a standard impedance level and signal level.

video connections to minimize static, interference, and ghosting. Video signals are especially subject to corruption, and great care must be taken with these connections.

NTSC baseband and RF modulated video signals can be easily connected over modest distances. Depending on the driver circuits used and the sensitivity of the circuitry on the receiving end, acceptable quality can be achieved at 100 to 200 feet, if good coaxial cable is used. However, RGB signals are much more limited, and cabling for them is normally contained within the same room. If significant distances are needed for RGB, you will have to provide booster amplifiers or else convert the signal to another form for transmission. Fiber optic adapters are a great solution for video signals, because the fiber cables can run fairly long distances with very little signal deterioration. The quality of the converters will have the most effect on video quality.

Life without Wires

It is certain that the home of the very near future will have an increased use of wireless technology. This will be true not only of our computer network connections using WLAN technology, but also of rather mundane connections, such as those for the computer keyboard and mouse. If you have never used a wireless mouse and keyboard, you really need to get one of those right now. You cannot imagine how limiting the cord on the mouse is until you get rid of it! The cord never seems to stretch far enough. It gets in the way. If you bunch up enough of the mouse cord on your desk to have plenty of range, the cord often moves the mouse when you let go. If you want to use your keyboard or mouse on a "busy" desk, you will have to route the cords around the other items while avoiding the mouse cord.

It is a lot more pleasant to get rid of the wires entirely. You can get a variety of styles of wireless keyboards and computer mice[11] (see the Tech Tip on Wireless or Cordless?). You can get a separate wireless keyboard and a wireless mouse. Both use simple wireless adapters that connect to your computer where the wired mouse and keyboard normally do. There are also versions that combine both functions into a PS2

[11]Maybe "mouses" fits better because we are not talking about rodents, but mice is the accepted term.

[12]USB keyboards are handy to use, but some PC BIOSs cannot recognize a USB keyboard. Most PCs that use Windows 98 or above will support a USB keyboard.

adapter or a USB adapter.[12] The adapters get power from the PC interface, so no external power transformer is needed. The wireless keyboard and mouse need internal batteries for power. It turns out that wireless keyboards and mice use very little power, so one set of batteries will probably last a couple of years.

Tech Tip: Wireless or Cordless?

The terms *wireless* and *cordless* have become interchangeable. In the keyboard or mouse application, the purpose of the connection without wires is to eliminate the keyboard or mouse *cord* or *cable*, which is rarely called a wire, so why is it "wireless" when it disappears? The term *wireless* is used often because it is a much more prevalent advertising term. Wireless is "hot" and cordless is not.

You can use both terms interchangeably when you mean a cord replacement. The only confusion results from comparing the very limited range of these adapters. It is very clear that you mean the short-range RF and infrared technologies when you say "cordless." However, to say "wireless" is ambiguous, unless the term is used with a defining word, such as keyboard. By the way, the same confusion exists between cordless phones (short-range devices that connect to a conventional telephone line within a location) and wireless phones (properly called cellular phones).

Now that some of the short-range technologies, such as Bluetooth, are being used to eliminate keyboard and mouse cords, the wireless/cordless duplicity has taken on a new life. Few users or manufacturers would refer to Bluetooth as cordless, although it is clearly not the classic longer-range wireless. The compromise seen most often in print is to call Bluetooth and the other competing technologies "short-range wireless."

Cordless PC accessories are available in RF and infrared versions. The infrared accessories use an invisible beam of light and are strictly for line-of-sight operation. That means that, if something such as your coffee cup gets in the way of the transmission between the keyboard or the mouse and the infrared adapter, the signal will be blocked. This may be a little disconcerting, causing your mouse pointer to move erratically or your keyboard to mistype (we never make keyboard mistakes, so it must be the connection). However, the infrared technology works very well and is impervious to the type of interference that happens with RF wireless connections.

RF cordless accessories use a short-range RF wireless connection medium. That is to say, there is method to this madness. Several frequency ranges are available for these unlicensed cordless accessories. In the United States, these devices operate under Part 15 rules of the FCC. Such devices are severely restricted in power output, and only certain frequency bands are used. The advantage to RF cordless operation is that absolute line of sight is not required. When used at short range, the RF adapter needs to be placed somewhere in the work area that allows communication to the accessory. Most of the time, this will be a desk or table near the host computer. Some of the early RF cordless accessories could choose between only two possible frequencies. This allowed a separate cordless mouse and cordless keyboard to be used on the same computer. Range is generally less than 10 feet (3 m), which is adequate for easy-chair surfing.

Bluetooth to the Rescue

The basic restriction for most cordless PC accessories is that only a limited number (such as two) can be used in the same area. In other words, most of these cordless/wireless devices operate on only one or two frequencies using conventional RF transmission methods. If you happen to have more than two cordless devices that use the same frequencies, they will interfere with each other. For example, one major brand of wireless keyboard has a two-position switch to set the operating frequency. If you also have a mouse that uses the same type of RF connection, you must set it to the other of the two frequencies. Then if someone who is nearby (in the next cubicle or in the apartment next door) gets one of these cordless devices, it may interfere with yours. As the worst case, you may find someone else typing on your computer...the man who wasn't there. Of course, you could inadvertently be doing the same thing on somebody else's cordless PC. This is not a very good way to meet neighbors.

What is needed is a way to have a moderate number of short-range cordless devices that are self-aware, so they can share the RF spectrum and not interfere.

A new wireless technology called Bluetooth[13] provides cordless devices with everything they need: short-range, low power, multiple

---

[13]Bluetooth™ is named after Harald Blaatand II, king of Denmark from 940 to 980 A.D. You can probably guess how *blaatand* translates into English. *Blaatand* was a nickname for the king who once ruled all of Denmark and portions of Norway. The rest is history.

devices sharing the same frequencies. The technology is targeted primarily for short-range groupings of up to seven active devices (see the Tech Tip on Bluetooth Networking). Bluetooth devices are characterized for three classes of power output. The basic level, Class 3, is the lowest power, at only 1 mW, or 0 dBm. At this power level, maximum range is approximately 30 feet (10 m), which is perfect for keyboards, mice, and printers. It also requires the lowest amount of circuit power, which is ideal for conserving battery power.

Bluetooth allows us to create a small, localized wireless network for our own private use. This type of network is aptly called a *personal area network* (PAN). A typical scenario for a Bluetooth PAN is illustrated in Figure 7.8. As you can see, this technology is great for "unwiring" your most common PC components and peripherals. You can also have an instant wireless connection between your PC and a Bluetooth-equipped cellular telephone or PDA. This would allow you to synchronize your calendar, your phone directory, memos, and e-mail, without having to make a direct connection.

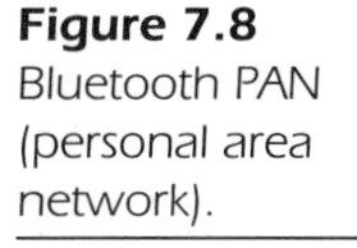

Figure 7.8 Bluetooth PAN (personal area network).

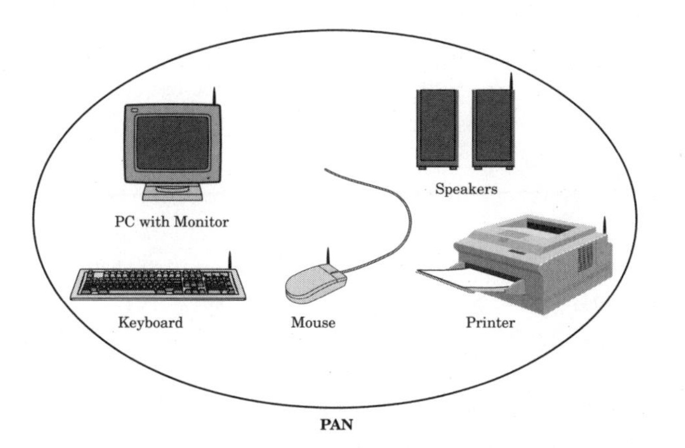

The basic level of RF power output for Bluetooth is 1 mW (0 dBm). The range expectations of 30 feet (10 m) are based on this output level. Bluetooth also allows two higher power output levels. Class 2 allows up to 2.5 mW, and Class 1 allows up to 100 mW. The higher output levels obviously allow more range between Bluetooth devices. However, the devices at both ends of a link must support the higher power level to realize the range extension. At the current time, most Bluetooth equipment is designed for the lowest Class 3 output power of 1 mW. In comparison, most IEEE 802.11 equipment has an output power of 20 to 100 mW (13 to 20 dBm), which yields a much larger range.

Tech Tip: Bluetooth Networking

Bluetooth uses the same 2.4 GHz frequency band as IEEE 802.11 WLANs. However, it uses FHSS for its primary mode of operation, in contrast to the DSSS of most 802.11 devices that operate at data rates of 11 Mbps and higher. FHSS is also allowed by 802.11, but it is not used much today because it is limited to 2 Mbps in most applications.

Bluetooth uses the standard 1-MHz channel spacing from 2.402 to 2.480 GHz, yielding 79 hop frequencies. Devices change hop frequencies 1600 times per second according to a set pattern. Unlike 802.11, which sets all devices to the same time/frequency sequence, Bluetooth allows two devices to transmit simultaneously on an offset pattern to create a full duplex link. Full duplex connections are required for applications such as cordless telephone links.

Several modes of operation are supported by the Bluetooth specification. Data transmission speeds of 57.6 to 723.2 kbps are allowed, as are three digitized voice channels at 64 kbps each. This data speed is well below the norm for 802.11 but very consistent with the needs for cordless device support. Even so, a 700+ kbps connection is not bad for Internet connectivity. The biggest restrictions on a Bluetooth network are the limited range and the limited number of active devices that may be interconnected.

Bluetooth devices connect to each other through a bonding process, which may include rigorous authentication and authorization. Even encryption is supported by some devices. However, it is unlikely that your Bluetooth keyboard or mouse would use these sophisticated techniques, so it might be possible to do (intentional) remote-keyboard havoc. You should also be aware of the potential security risk that could come from transmitting your keystrokes directly into the airwaves.

Authorized and bonded Bluetooth devices can form a small wireless network. A type of ad hoc network, called a *piconet*, connects a group of up to eight active devices (compared with 256 for 802.11). One of the Bluetooth devices becomes the master and up to seven other slave devices may actively join in. In addition to the eight active piconet devices, other devices can secure attachment rights while remaining "parked." A total of 256 devices may be parked on a piconet. Don't you wish they had this much parking everywhere? This gets better.

You can have multiple piconets in the same physical space, if the coverage areas of two or more piconets overlap. Devices in one piconet can potentially communicate with those in another piconet. The entire set of piconets is referred to as a *scatternet*. (Told you it would get better.) The device in one piconet simply uses a timeslot in the other piconet, when available. At least this allows sort of a moderated *detente* to occur, which minimizes interference.

Speaking of interference, one of the biggest problems in using IEEE 802.11 WLANs and Bluetooth devices in the same area is that they may interfere with each other. Unlike the mutual awareness of the scatternet, these technologies compete for the same spectrum in the 2.4-GHz band, and they treat each other as interference sources. As you might expect, this can cause real problems. In a theoretical open area, with a matrix of 802.11 and Bluetooth devices, the 802.11 network suffers more interference issues than the Bluetooth devices, based on analysis contained in IEEE and Bluetooth Special Interest Group white papers. In general, a W-NIC must be within 10 to 20 m of an AP to be comfortably resistant to interference from a nearby Bluetooth device. As a result, some large-scale WLAN installations have placed a moratorium on Bluetooth and other 2.4-GHz spectrum devices, such as certain cordless phones. Efforts are underway to provide a mechanism for Bluetooth and 802.11 devices to coexist. Although there may be some performance trade-offs, peaceful coexistence is crucial in the increasingly crowded 2.4-GHz band.

HomeRF

In addition to Bluetooth, another technology called HomeRF has invaded the 2.4-GHz space. This technology is intended primarily for audio and telephone cable replacement, but, as with everything else, it also supports data networking. As a matter of fact, HomeRF Version 2 bumps the base rate from 1.6 Mbps to 10 Mbps and supports voice, data, video, and multimedia streaming. Video and multimedia are particularly hungry for bandwidth, and the 2.4-GHz band is very finite. It remains to be seen how much of a force HomeRF will be, but with different categories of vendors supporting technologies that compete for the same wireless spectrum, you can expect an interesting fight.

The Totally Automated Home

The final step in the high tech home is total home automation. How would you like to be able to control your lights, heating and cooling, provide interconnected home security features, and even interconnect your appliances? These capabilities and more are coming soon to a home near you! Perhaps they will invade your home, too.

Control Your Lights

If the concept of an automated home is a stretch, you might consider that we have been able to use remote controls to turn lights off and on for many years. A lighting-control system called X-10 that uses household electrical wiring for signaling between a controller and the light switches has been around for at least two decades. This type of signaling is properly called *carrier-current communication* or *power-line carrier* (PLC) *communications*.

In classic PLC signaling, the control or communications signals are placed on a carrier in the range of 100 to 300 kHz. These frequencies are well above human hearing and out of the range of most home electronic devices. In addition, the PLC frequencies are above the rapidly decreasing harmonics of the 60 Hz (or 50 Hz) AC voltage that powers our lighting and appliances. PLC signaling is designed to drive into the very small (and variable) impedance of the household power system, and adequate signal strengths can usually be achieved throughout the house.

A wide variety of lighting and appliance controls is available. The controls can be divided into three categories:

- Variable-intensity controls (dimmers)
- Inductive-load variable controls (fan controls)
- Relay-closure controls (appliance controls)

The variable-intensity controls are perfect for dimming incandescent lights. Fan controls can change fan speed for perfect comfort. The relay controls are used for large inductive loads (such as appliances with motors) and fluorescent lights. Special dimmer controls are also available for fluorescent lights equipped with compatible dimming ballasts. A special category of the relay control is available to close contacts, for example, on a security alarm circuit.

Several competing systems are being marketed that use the power line for communication. The original system that made use of PLC is the X-10 system. This system has a simple ID mechanism that allows up to 16 "house" codes and up to 16 "unit" codes, for a total of 256 combinations. Because PLC signals can theoretically travel over the power lines to the next house, you could have some real fun with the neighbor's lights. Seriously, this 16-unit limit makes whole-home automation much more difficult. Of course, you can use more than one controller, with each set to a different house code, but most controllers are limited to 8 or 16 devices maximum. In addition to the manual controllers, PC interfaces and software are available to computerize your system of controls. You can get pretty sophisticated with this, but the system is still fairly limited in terms of numbers of devices and type of communication.

Another system that is really gaining favor is the CEBus system (see the Tech Tip on CEBus, ANSI/EIA-600, and HomePlug). This system allows many devices to be controlled over PLC. CEBus uses an advanced form of spread spectrum that minimizes the effect of interfering signals. CEBus operates in the same 100- to 400-kHz frequency range used by other remote devices (including baby monitors and intercoms), so it is subject to some potent interference sources.

With the same advanced spread spectrum system, HomePlug ups the ante to 14 Mbps by using frequencies of 4.5 to 21 MHz. This allows it to offer data rates comparable to 10BaseT Ethernet and 802.11b WLAN without additional wiring. In addition, there is plenty of extra bandwidth to add voice and limited video.

Tech Tip: CEBus, ANSI/EIA-600, and HomePlug

CEBus is a home automation standard that has been emerging over the past decade. This standard is now officially part of the national and international landscape and is known as ANSI/EIA-600. Products are now being offered that meet the general specifications of CEBus. Fortunately, CEBus is one home technology that does not use the 2.4-GHz WLAN band, so it will not cause interference with 802.11 networks.

CEBus uses a type of spread spectrum approach that is fundamentally different from FHSS and DSSS. The formal name for this method is the *swept-frequency pulse-chirp spread spectrum system*, but it is commonly called the *chirp system*. Basically, the chirp system sends a signal that changes continuously over a range of frequencies. In essence, it uses infinitely small frequency hops as the signal energy is swept through the signal range. If you could listen to the signal on a standard receiver, you would hear a "chirp" as the signal swept past your receiver frequency.

A chirp spread spectrum system has FHSS's resistance to a constant, jamming signal, and also has the 10 to 14 dB processing gain of a DSSS signal. It is also very simple to transmit and receive. These characteristics are ideal for simple home networking and control, with the power line as the communications cable. The chirp signal has a very good resistance to the noise sources that are present in that environment and provides additional performance needed to support sophisticated applications.

Currently (no pun intended), CEBus and HomePlug chirp-style systems are available in frequency ranges of 100 to 400 kHz and 4.5 to 21 MHz, respectively. The lower range is suitable for simple controls and data interfaces to 9600 bps, and the higher range can support data rates up to 14 Mbps.

To complete your lighting control system, you can consider one of the new RF lighting controls. These systems operate at 418 MHz (416 MHz in some locations) and thus present no interference to PLC or WLAN systems. The control distance is limited to a radius of about 30 feet, but the signal can be repeated up to four times. A 120-foot reach should accommodate most homes, unless you are king of the software business. A wide variety of preprogrammed "moods" are available for these controls. Remote controllers have several buttons to set the mood. Touching the button defines the mood of the moment, such as Evening, Entertain, Reading, Theater, or Bright. Instantly, your desires are communicated to a set of lighting controls for the room, and the perfect lighting environment is set.

A typical lighting control with mood settings is shown in Figure 7.9. On most controls, the settings are clearly labeled. You can also control individual lights through their dimmer switches. One of the important features of a home lighting automation system is that you can control your lighting in a variety of ways, even manually.

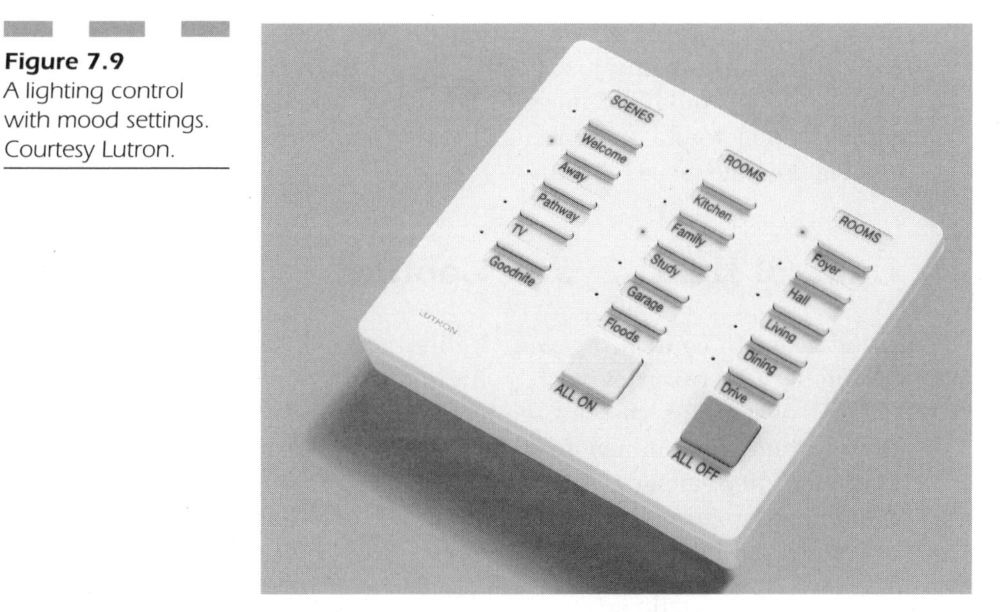

Figure 7.9
A lighting control with mood settings. Courtesy Lutron.

There are several similar types of lighting control systems. Most systems have a variety of controller options. Some controllers have as few as four control buttons, and some have as many as 20. Controllers may control the attached lighting devices individually or as a group. For the mood lighting systems, wireless controllers are available that have the room's mood settings programmed. Depending on the specific system you get, you can use wired, RF, or infrared remote controls.

Many computerized controllers are available to work with these systems. Most systems have regular computer interfaces and specialized control software. Several options are available for dedicated computerized controllers. These controllers are self-contained but give you the full functionality of a computerized unit.

You can also control your home lights from your car visor. How convenient it would be to be able to control the home lighting settings as you open your garage door. The system could turn on the lights along the driveway or the footpath to light your way. It could also brighten the door lights, turn on the hall lights, and even start the music, if you have it programmed to do so. What a life!

Some systems allow remote access through a telephone or a data connection. You could be in Maui (here we go again) and check on your home through the Internet. Just call up your computerized controller and view your security cameras. Bring the lights up to brightly light the areas you want to view. Then program your house system to automatically turn on the lights just before you get in from your flight home. Of course, if you would just move to Maui, you wouldn't have this problem, although you would still probably want your automated home lighting system. For the mood, of course!

Control Heating and Cooling

In addition to controlling your lights, you can use home automation technology to interact with your heating, ventilation, and cooling (HVAC) system. As you can see in Figure 7.10, a computerized automation system can connect to a computerized thermostat and temperature/humidity sensors. As a matter of fact, in a complex system, you could have two or more thermostats and several temperature sensors.

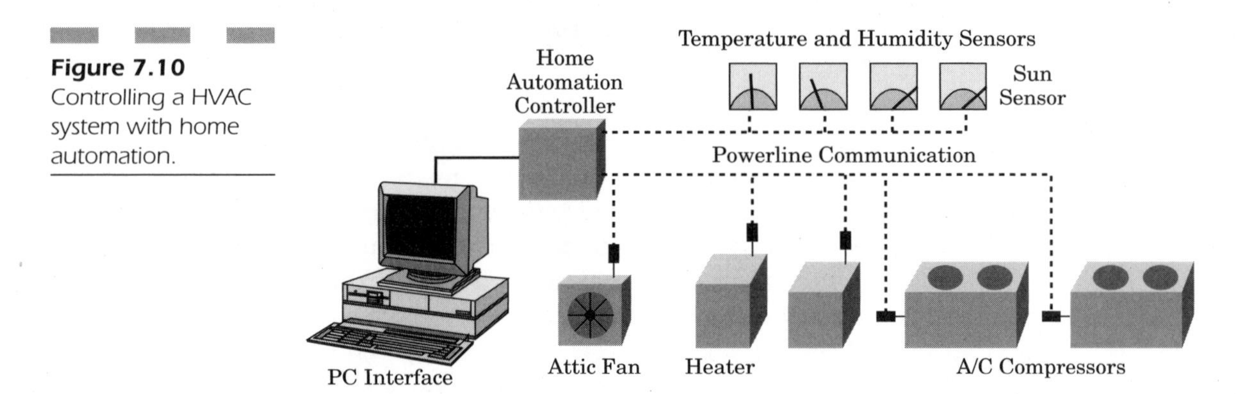

Figure 7.10 Controlling a HVAC system with home automation.

The main purpose of a HVAC system is the comfort of the occupants. As you are aware, there is a need for heating and cooling changes throughout the day, as outside temperatures change. Many factors are involved in determining the proper levels of HVAC use for your comfort. The ideal temperature can change with the outside temperature and the relative humidity. It can also change with the amount of sunlight and your own activity level.

A computerized controller can compensate for these variables by modifying the heating or cooling cycles. If you have more than one HVAC unit, their operation can be coordinated and adjusted for time of day, amount of sun loading, humidity, and other factors. For example, if you are using summer cooling in high humidity, longer running cycles will pull more humidity out of the air, thereby increasing your comfort. A computerized system can also adjust for a variety of daily and weekly schedules, to increase your energy efficiency by not operating the HVAC excessively when you are at work or away from the home.

You can use the same computerized system to set the lights to different levels during the day and night hours. If you always want the outside lights on at dusk and off at dawn, you can use a solar sensor connected to your home automation system to take care of this detail. You may also want to turn on an attic fan or a dehumidifier when the conditions are right.

Only some of the so-called home automation systems can control your HVAC system. In addition, you must have a controller that imitates traditional controls, such as a thermostat, or an HVAC system that is set up for computerized control. (As the psychologist would say, it must want to change.) Many of the simple lighting control systems have only primitive HVAC controls, or they have no such controls.

Provide Home Security

Many of the more sophisticated home automation systems have security functions. These can provide some built-in security capabilities, or they can interface with third-party systems. There are definitely some tradeoffs to be aware of. You may not be able to get the best of both worlds.

The methods to interface with a third-party system include Ethernet, RS-232 Serial, and contact closure. If the systems connect via Ethernet, they will have to support a common control language. RS-232 commands will be fairly straightforward, and the home automation system will have to adapt to the control commands of the security system. Contact closure is the simplest method but also the most limited. With contact closure signaling, you might be able to have the security system command the lighting system to turn on all the perimeter lights when the alarm is tripped. Conversely, you could have certain events under the watchful eye of the home automation system communicate with the security system.

To do something sophisticated, you will need advanced systems that communicate over Ethernet or over a PLC connection. For example, you might want to check the status of your security system from a remote location. You would use the remote communications capability of the home automation system to report the status of the security system, perhaps on a Web page.

Network Your Appliances

More appliances are being offered with interfaces to home automation systems. It is now feasible to have a refrigerator, stove, or even coffee pot that talks on the home network. You could use the built-in browser display on the front of the refrigerator to order more food from the Web grocer, or you could control your oven from the home automation system. A sophisticated coffee pot could allow you to change the start time from the bedroom, after you decided you didn't really want to get up that early.

The most logical way for appliances to communicate with the home automation controller is over the very power line that connects them to AC voltage. One of the PLC technologies, such as CEBus, is perfect for this. It already supports digital control transmission, so it is well adapted for appliance control. More and more appliances will become available with automation interfaces. Soon we will be able to have our homes totally automated, with all the attendant benefits.

All these things and more are on the horizon. Home automation is the key. Once you have a sophisticated and general-purpose home automation controller installed, you can automate almost anything. We just hope the appliances don't begin talking about us behind our backs!

CHAPTER 8

Wireless Office and Campus LANs

Chapter Highlights

- Large Wireless Networks
- Locating APs
- Avoiding Barriers to Wireless
- Wireless Campuses
- Wireless Bridges
- Project: A Large Office/Campus WLAN

We have covered relatively small home and office networks up to this point. This chapter is about building large WLAN systems. We will use the knowledge we gained from building smaller WLANs as a stepping-stone to larger networks.

WLAN technology has wide application for commercial use. One of the biggest barriers to effective and efficient conventional computer networks is the inflexibility of the network infrastructure. The infrastructure wiring, or the cable plant as it is called, is a highly defined and fixed part of the network architecture.

Not only is the wiring fixed in place, anchored to the walls and cubicles by network outlet boxes, but its performance is set in stone at the time it is installed. If your network cabling becomes inadequate for your needs, you can no sooner push it to the next performance level than you could run a gasoline automobile on electricity. You must replace the whole infrastructure to upgrade. Replacing that infrastructure is a very expensive proposition, as those of us who have had to upgrade from 10-Mbps Category 3 cable to 100/1000-Mbps Category 5e cable have found out. We now face yet more expensive and disruptive upgrades to other technologies, including new Category 6 and 7 wiring systems.

In addition, conventional network wiring is very inflexible. If you need a network outlet where you have none, you must pay to bring in another line. If you need two outlets where you have one, you are faced with cumbersome additional hardware or wiring an additional line. If you have a large area that does not lend itself to cabling, you have no aesthetically pleasing way to bring in the cable. In addition, temporary network connections are often handled with a dangerous and obtrusive cable strung across the floor. Conference rooms and multiuse areas, such as company cafeterias or lobbies, also present problems for network connectivity.

Many authorities estimate that network wiring infrastructure costs have grown almost equal to the costs of computing hardware, excluding large servers. This means that you can expect to spend as much on your cabling and connectivity as you do for your PCs and software.

WLANs can be used to supplement this wired network. WLANs are particularly useful in providing network coverage for meeting rooms and common areas. Many private and public organizations have found that WLANs are, financially, a viable alternative to conventional network cabling. This trend is gaining acceptance in many companies and campuses.

Large Wireless Networks

Large networks always require additional considerations in planning, design, and implementation. In a smaller wireless network, such as the one you built for your home or small office, planning and design are minimal. Essentially, you had to count up how many W-NICs you needed, add any PCI adapters, take a good guess at where to locate the AP bridge, and install everything. You spent most of your time on the implementation phase, installing and configuring the W-NICs and AP.

This minimal planning and design, however, will not allow you to produce a successful large WLAN network. In essence, the bulk of the task shifts from that of a simple network, where preparation is minimal and implementation modest. A large network requires adequate design of the wireless access nodes to ensure adequate coverage for the users. In a typical office building, you have to be concerned about providing placement of multiple APs to gain proper wireless access. You have to locate and plan for power, network connections, and mountings for each AP. In addition, you frequently have to provide external antennas to accommodate building construction and arrange the placement to gain the wireless coverage you need.

Building Complex Networks

The key to building a successful complex WLAN network is to couple adequate site planning with actual field measurements. This is an important process that is often overlooked. Until you have gained a great deal of experience in situating APs, you are unlikely to be able to make the highly educated estimates that are required to shortcut these two steps.

Site planning involves determining the likely locations for APs, based on the desired coverage areas in the building or campus. If at all possible, you should obtain site and building drawings for your proposed installation. These drawings will allow you to determine the logical placements for APs and sketch where the coverage areas for each AP will be. Although you cannot see through walls, APs can, or rather their wireless transmissions can, and you will want to know where those areas of the building are that can be covered through this wireless equivalent to x-ray vision.

At the same time, a strong AP signal can have unwanted effects on your wireless network's operation. For example, a W-NIC wireless station will often look for the best signal available from an AP. If you want stations in a particular room to access the subnet serving that room, you will want them to connect to an AP that is connected to that same subnet. However, an AP in an adjacent room or hallway may produce a stronger signal. A station could inadvertently connect to that unintended AP.

The field measurement component of your design planning is the "proof of the pudding." After you have done your walk-through and your paper planning with the building drawings, you should test your assumptions. What does that mean? You didn't make any assumptions? Sure you did! You assumed several things, including the integrity of the building drawings.

Some of the common assumptions that we all must make in planning an office WLAN installation include acceptable AP placement, power availability, antenna type, and signal range. You may think that you can place an AP high up on that foyer wall until you find that there is no nearby power or network cable connection and the building manager will put a contract on you if you mess up the appearance of the area. Corporate image often means more than computer pragmatism, and politics and aesthetics will almost always win out.

Did we say you needed a wired network connection, too? And you thought this was a wireless network. Obviously the purpose of the wireless network is to connect into our wired company network, so you must have a wired connection for every AP. Besides, APs don't do a very good job of talking to each other,[1] so they need a wired backbone to interconnect. This means that you need a great location, A/C power, *and* a 10/100BaseT network connection. These factors make planning your wireless network much more interesting, don't they?

Now, you need to do the actual field measurements. There are two ways to accomplish this: *direct link testing* and *field-strength measurements*. If you have a simple installation requirement and budget constraints are a problem, you should probably use simple testing with an actual WLAN client by running the link test utility. The better link test routines include a bar graph and the numeric values of the wireless link from both the W-NIC's standpoint (at the laptop) and the AP end of the link. This will probably require you to run the AP management utility because most W-NIC utilities report only whether the link is good or bad from the W-NICs perspective. That would mean that you would know

---

[1]APs can connect to each other to bridge distances wirelessly, but in general, they cannot act simultaneously as an infrastructure hub and wireless-to-wired network bridge.

whether the W-NIC was receiving adequate signal from the AP, but not vice versa.

Unless you are a radio wave, you cannot predict how the WLAN signals will propagate through solid objects, such as windows, walls, doors, and floors. You would be surprised how many walls and floors hide metal air ducts and pipes. In addition, many modern buildings have interior walls that are constructed with steel structural members as opposed to wooden ones. These metal studs, rafters, stringers, and joists can contribute to coverage problems for wireless networks. Even wall coverings can be metallic or foil covered, which can cause significant signal loss for your WLAN.

To accomplish an accurate direct link test, you must provide a temporary mounting for the AP or its external antenna at or very near the desired point of installation. An easy way to do this is to mount the antenna to a pole or a board of sufficient length to temporarily test the area of the wall or ceiling where you expect to permanently mount it. While one person holds the antenna in place, the other can take a laptop with an installed W-NIC to the locations where coverage is desired and note the signal strength on the AP management utility. If the wireless station is to be placed on a desk or a table, move the laptop to that approximate position and note the link status. Record all the values on a chart, so you can later transfer the measurements to the site drawings.

If you encounter a value that is obviously too low, you might take an external antenna to connect to the laptop's W-NIC to see if you can get an adequate improvement. If there are very few locations that experience a low signal problem, it may be less expensive to provide a booster antenna than to install another AP. Check our Web site at www.buildyourownwirelesslan.com for more information and links on these signal booster antennas.

In addition to the direct link test, it is possible to do a very thorough and unbiased test using portable field-strength measuring equipment. Actual field strength tests are considered far superior in predicting WLAN performance over a test area. An RF source may be coupled to the actual antenna to be used. The source simulates an AP. Field-strength measurements of the EM field are then made at numerous points in the area where coverage is desired. The measurements are made with highly calibrated antennas and sensitive RF voltmeters tuned for the WLAN band to be used (e.g., 2.4 GHz). The signal strengths are recorded and then are transferred to the site plan or building drawings. Finally, the regions where adequate signal strength was detected are plotted within the measurement field (Figure 8.1).

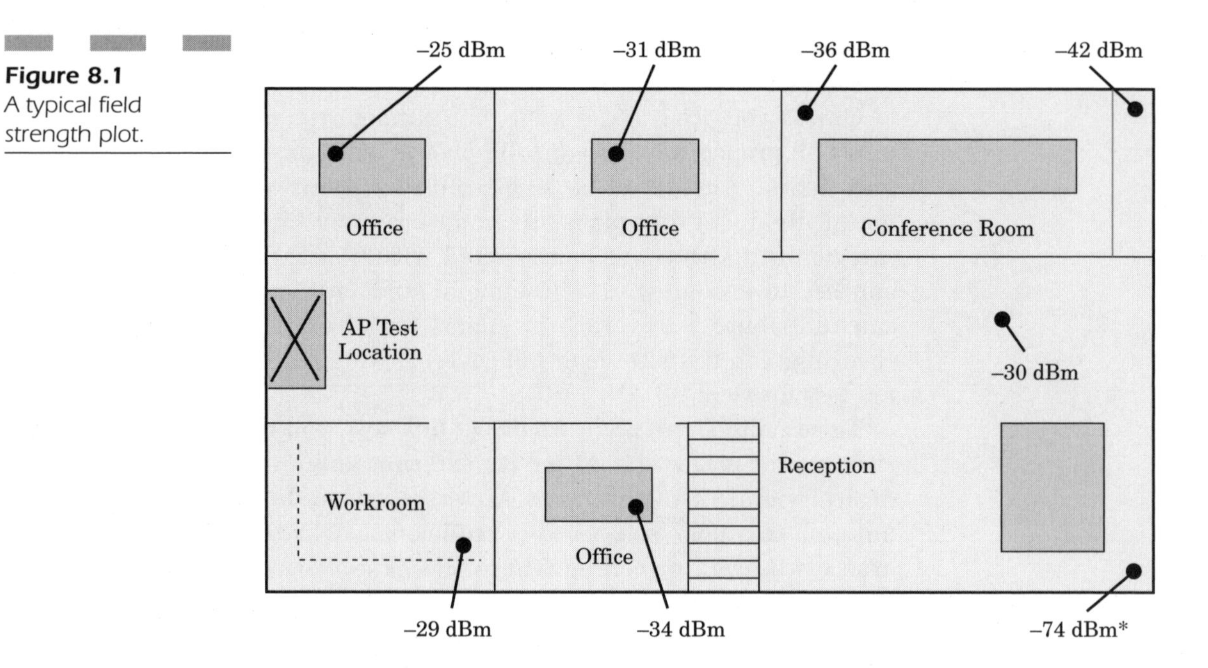

Figure 8.1 A typical field strength plot.

The signal strength required for a link changes with the data rate that is being supported. As you saw from the information in Chapter 4, a WLAN's range is related to the maximum data rate on the link. What the rate/distance chart really indicates is the amount of signal strength needed for wireless operation at different speeds. In other words, it requires more signal strength to connect at a higher rate, with distances being equal. The corollary is that longer distances are supported at lower data rates.

As the WLAN system designer, you must determine what data rates are acceptable for your users. If a typical WLAN user can tolerate a 2-Mbps data rate, then you may be able to place the APs less densely. Conversely, if a guaranteed rate of 11 Mbps is needed, under most circumstances, you will have to place the APs more densely.

The field measurements, whether carried out by the direct link test or by field-strength testing, will be your confirmation that the AP locations you have predetermined are adequate. It will also quickly reveal the weak points in your coverage.

There is a pay-as-you go method of AP site placement that you may want to use. If your requirements are not too strict, but you definitely want to provide good coverage on a tight budget, you could use this method. Start with a relatively sparse placement of APs. Use link testing to verify that your basic coverage assumptions are accurate. Allow locations where secondary APs can easily be added later, if a coverage hole is found. Make it clear to management that the method will not guarantee complete, high-speed coverage everywhere, but that you can always add additional AP nodes to boost performance selectively.

Channel Reuse

Another item you will need to plan for in a complex network is channel reuse. To cover a large building or open area, you will need to have many APs. You can consider the WLAN APs in a building to be similar to a cellular grid. The aim of your design is to place APs in such a way that a wireless station is always within the coverage area of at least one of the APs. So that they will not interfere, adjacent APs will need to be set to different frequencies. Figure 8.2 shows a typical channel reuse scheme.

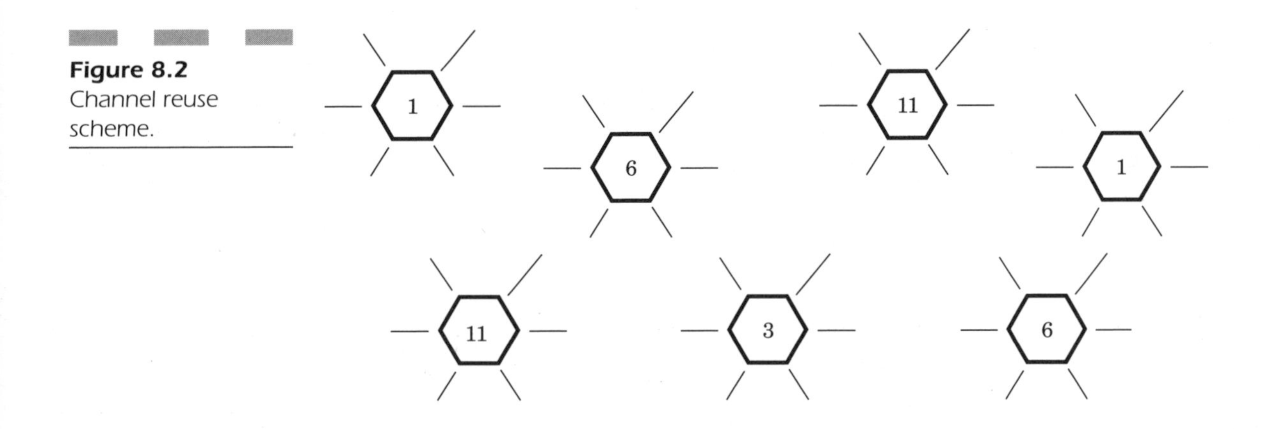

Figure 8.2 Channel reuse scheme.

There are 11 channels available for use in North America.[2] Because of the potential for adjacent-channel interference, it is customary to separate the frequencies of neighboring APs by three to six channels. This is particularly important at networks running at the higher speeds

[2]The IEEE 802.11 standard specifies a band plan of 14 channels in the 2.4-GHz band, 11 of which are used in the United States and Canada. See the Appendix for a complete listing.

Tech Tip: In-Band Interference

RF engineers have developed some very useful terms for the common types of interference to communications links. All RF receivers have a response in the frequency domain called *selectivity*. Receivers also have a minimum received signal level, called the *sensitivity*, that yields an acceptable *signal-to-noise ratio* (and thereby a usable signal). The selectivity of a receiver is similar to a bell-shaped curve, with really steep sides. At microwave frequencies, it becomes increasingly difficult to sharpen the selectivity of a receiver. Consequently, receivers, such as those used for 2.4-GHz WLANs, cannot reject strong, nearby signals.

An interfering signal on exactly the same frequency that your receiver is tuned to is called *cochannel* interference (on the same channel). Interference on a nearby frequency is called *adjacent-channel* interference. Interference that results from one signal mixing with another very strong signal (that normally would have been rejected by the RF or IF stages of the receiver) is called *intermodulation*.

Receivers are subject to intentional and unintentional interference sources. An example of an intentional source is another AP on a nearby channel. An unintentional source might be a microwave oven or another legitimate user of the 2.4-GHz band. To reduce interference from another AP, you should try to separate your WLAN receiver from the source by distance or frequency. What this means is that you can effectively move the interfering device down on the receiver's response curve by increasing the distance between you and the device or changing the operating frequency of one or the other device.

In simple terms, you should avoid using W-NICs or APs that are operating on the same or nearby frequencies because they will cause interference to your desired channel frequency. In addition, you should take steps to avoid non-802.11 interfering sources, such as BlueTooth, HomeRF, or 2.4-GHz cordless telephones, because they can produce co- and adjacent-channel interference and occasionally intermodulation interference.

because they are more sensitive to levels of interference that would be ignored at lower speeds. There is more information on adjacent- and cochannel interference in the Tech Tip on In-Band Interference.

A multistory building presents an interesting problem for planning a WLAN and channel reuse. You have to consider the fact that the wireless signals most certainly will penetrate floors and ceilings, at least to some extent. Even if the building floors are cement with structural steel, they may allow substantial penetration of the WLAN signals, although there may be an accompanying reduction in signal strength. Even so, in most buildings, this signal leak-through may be used to your advantage, as we will see.

From a channel reuse standpoint, signal leakage through the walls and floors of a building interfere with the level-ground ideal of a cell structure. You will have to play a bit of three-dimensional chess to take this into account. Not only will you have to avoid the adjacent channels on nearby APs on the same building floor, but you will also have to watch out for the APs on the floor above and the floor below.

Linking Multiple Buildings

APs are, by their very nature bridges. They were originally designed to bridge a wireless infrastructure and a wired network backbone. But what if there were only one wireless station connected, and what if that wireless station were another AP? Why couldn't the two APs simply carry traffic between their two networks? Well, they can, but to properly implement this *wireless bridge link*, they need to run special software.

The reason for this is simple. The bridges expect to connect to W-NICs that support only one IP address. The AP acts as a local bridge between those W-NICs and a wired network. To act as a wireless link, the two bridges must operate in a mode that is very similar to a WAN bridge (Figure 8.3). WAN bridges have been around for a long time, although we are more familiar with WAN routers today. A Layer 2 bridge learns the MAC addresses on each side and passes only packets destined for the opposite side. WAN bridges use a protocol between them that encapsulates Ethernet packets. Leased-line bridges use protocols like HDLC, whereas our wireless bridge uses a variation of IEEE 802.11 for the link.

This is a dandy way to get a network connection between two buildings. If cabling is a problem or impossible, a wireless bridge may be just the thing. It is even possible to transport a network connection to a distant building. Most APs do not come with bridging software, but many manufacturers offer it as an option. Standard protocols are under development to support this type of link. However, for the present time, it is prudent to use the same brand of bridges at each end.

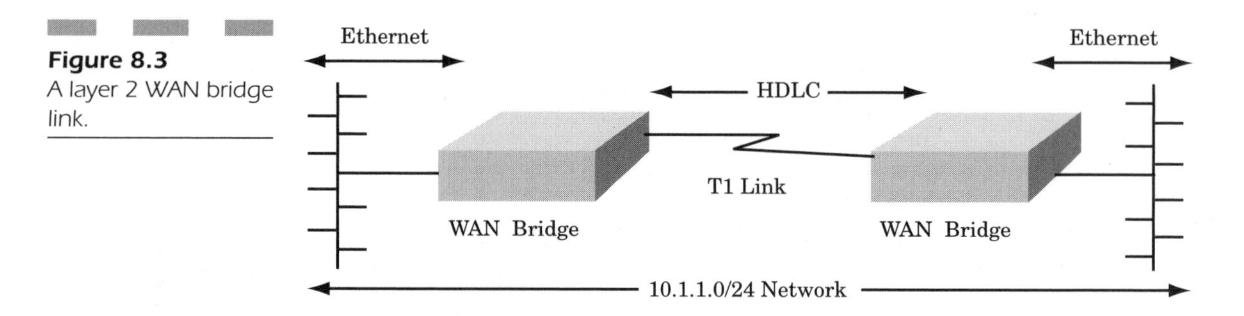

Figure 8.3
A layer 2 WAN bridge link.

There are several reasons to use a wireless bridge instead of conventional cable to connect two buildings. A wireless bridge offers a fairly inexpensive solution for network extension. Because the link end-points use ordinary unlicensed WLAN hardware, they are very simple to set up. In addition, they offer a link speed from 11 to 54 Mbps (with the new hardware standards). An 802.11 wireless bridge link can be set up for less than $1000, with no special authorizations. In comparison, a standard, microwave link costs around $15–20,000, may operate at rates as slow as 1.5 Mbps, and may require government licensing.

A typical application for a wireless link is a situation where regular network cabling, such as a fiber optic cable link, is impractical, expensive, or impossible. For example, if a busy street, an expressway, or even a river separates the two buildings, it may not be possible to run your own cable. Many cities have restrictions against running private cabling across any street or alleyway. If the buildings are several blocks apart but have line of sight, it might be feasible to use a wireless link.

Another reason you may want to use a wireless link is the higher speed. Conventional wide area lines, leased from the local exchange carrier (the phone company), use copper cable pairs for data transmission. These links are limited to T1 speeds of 1.5 Mbps. In addition, the T1 bridges and interface equipment are expensive, costing as much as $2000 per end, and you will be charged a monthly line fee of $500–1500. For the first year of operation, this could cost more than $20,000! Higher speed links are available, but the termination equipment and the line charges are higher. Line charges for a T3 line, which has a rate of 45 Mbps, may run 5 to 10 times that of a T1 line.

Speed of installation may be another good reason to use a wireless bridge. Once you have planned the link and purchased the equipment, you can install the bridges immediately. There is no waiting for the phone company (typically 30 days), a federal license (typically 90 days), or professional installation. Unless the link is distant, you will only need

Yagi[3] antennas, which have a 15° beamwidth and are not difficult to aim. This quick setup also makes the wireless bridge ideal for temporary connections or for disaster recovery.

The wireless bridge link is built by installing two APs with two external antennas and making the wired network connections at each end. This is illustrated in Figure 8.4. The APs must be running appropriate wireless bridge software instead of their regular WLAN software. Wireless bridges can operate in a point-to-point mode or a point-to-multipoint mode. If you are trying to support wireless links to two or more remote buildings, you will need to use multipoint mode. Keep in mind that all of the remote links must share the bandwidth with the central wireless bridge.

Once the bridge software is installed, you can configure the two bridges for the wireless bridging, configure both for the same SSID, set the encryption level, choose the channel of operation, select the maximum link speed, and set the other parameters.

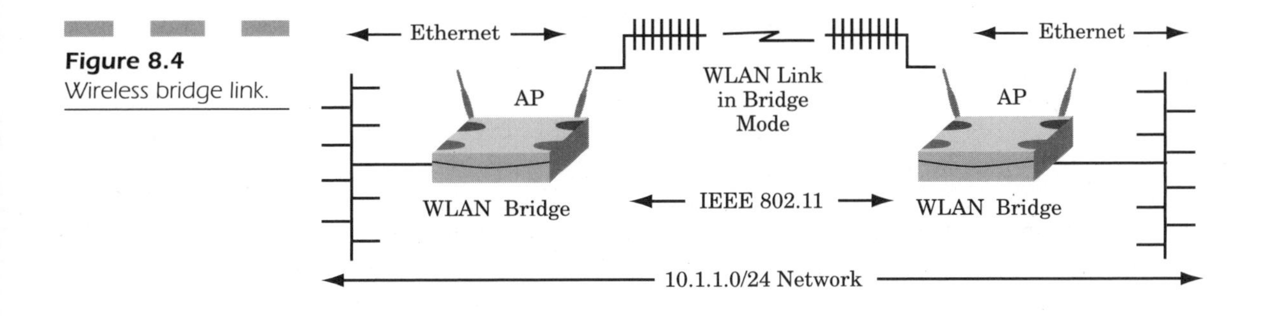

Figure 8.4
Wireless bridge link.

The management utility should let you see the signal strength provided by the link. Be sure to note the received signal at both ends of the link. If you are in a situation where the link is at or near its limits of signal strength for the speed of operation, you may need to increase antenna gain, add booster amplifiers, or lower the maximum link speed. Lowering the link speed may increase throughput in a marginal link. The WLAN hardware will always try to connect at the highest possible speed. If operation at that speed is marginal, the link may bounce up

[3]The Yagi antenna, more properly called a Yagi-Uda array, is a directional antenna. It consists of a driven element and a reflector element in parallel with one or more additional elements, called *directors*, also in parallel. The directivity is toward the directors, and the gain increases as more directors are added. Yagis for wireless networking typically have as many as 14 directors and produce a 12–15 dBi gain.

and down in speed, losing data packets along the way. This causes the connecting protocols (TCP/IP in most cases) to retransmit the lost data, which increases the traffic on the line and decreases throughput.

To manage the AP bridge links remotely, you will need to set an IP address for each. If you have not already done this, set each IP to an available address within the subnet at the host end. Because this is a bridge, not a router, both networks must be on the same subnet. If you need to set the remote network to a separate subnet for some reason, you will need to provide a local router external to the bridge.

Authenticating Access

With a large wireless network, you are more likely to be aware of the need to protect access than when you built a small wireless network for your home or small office. Either network is vulnerable to a wireless break-in, but a corporate network may be a more valuable target.

Basically, anyone with a W-NIC can access an open WLAN system simply by booting up a wireless card. Wireless signals do not contain themselves inside company walls. That means that anyone in your lobby, in the business on an adjoining floor, or even in the parking lot can get into your network. A WLAN is not like a wired one, where you theoretically have control over all of the cable outlets that can be used to gain entry to your network. The wireless signals go everywhere within range of the APs. Unless you intend to have a completely open WLAN system, as in a freenet, or a public WLAN, you will need to implement access control.

The most straightforward approach to access control involves authentication of the user. We covered much of the theory and practice of authentication in Chapter 6. The basic idea is that a potential user is restricted to an authentication server until the user is granted access. The user may need to enter an acceptable user ID and password, enter a secure token, or have a MAC address on the access control list. In addition, the user may need to have the proper encryption key to set up an encrypted or virtual private link.

A typical authentication scheme is shown in Figure 8.5. When the new wireless user tries to connect, the user's TCP/IP stack will request an IP address through DHCP protocol. The DHCP server initially assigns an address on a restricted subnet that has access only to the authentication server. The authentication server performs its duty, and if the prospective user passes muster, it assigns a new IP address that allows access to the network. That is a very simplified view of what occurs. Many methods exist to ensure proper authorization, and you can choose the method that best suits your overall network needs.

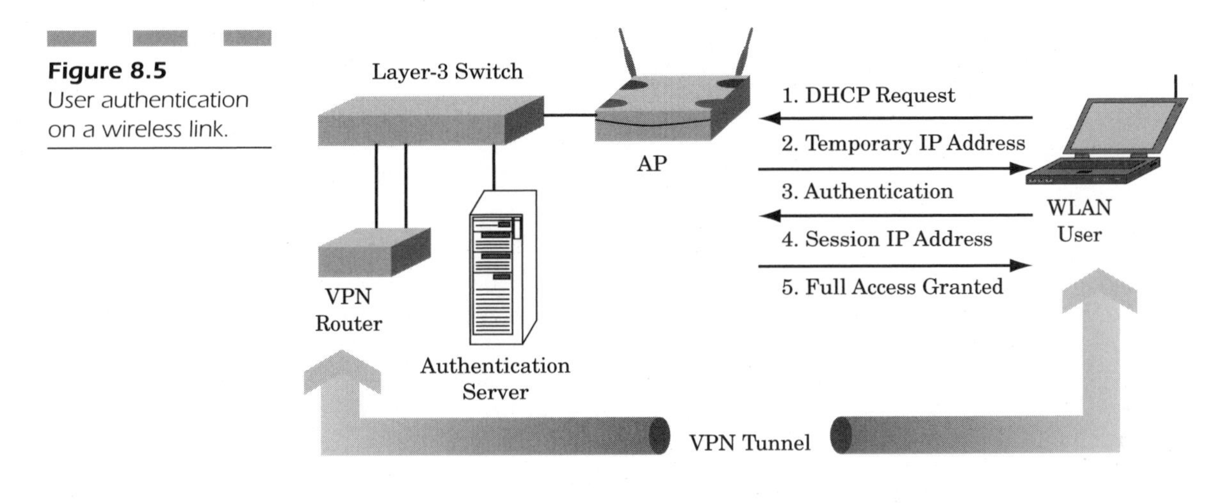

Figure 8.5 User authentication on a wireless link.

In addition to authentication, access may be effectively controlled through VPN encryption and additional firewalls. You must keep in mind that a WLAN is potentially an open door into the data domain of your company. You must at least lock the door and guard your network by controlling who gets in.

You might wonder why you would need a firewall inside your network. Firewalls are usually placed at connections to the outside world, for example, the Internet. Very simply, you do not have physical control over the air waves. So a WLAN is indeed a connection to the outside world. A corporate spy or even a casual WLAN user might gain access to your network. Imagine your surprise when sitting in that coffee shop, if your laptop computer brings up an AP at the bank across the street! Don't laugh, this type of thing has happened, and it is not all that rare.

A firewall with a VPN will help secure your network and your WLAN. This structure will secure your data transfer and access to your wired network. The firewall adds extra protection to make it more difficult for anyone to get all the way into your wired network. In fact, a number of organizations are taking this step to isolate the WLAN network from their corporate backbone. It can then be treated as a remote connection, with a higher level of security and access control than for the directly connected users.

Most large corporate networks have a fairly elaborate authentication and access control system already in place. It is a fairly simple matter to add WLAN nodes to that system. If additional router and firewall/VPN services are necessary, it may be a minor change to the system you are already using. Many Ethernet switches are also simple Layer 3 routers that can support an almost unlimited number of subnets and users. Likewise, many VPN/firewalls can support more than a single VPN, so you could provide the WLAN with its own, separate VPN.

A danger of inadvertent WLAN access comes from the ability of any employee or manager to install a simple WLAN starter kit and connect it into the corporate backbone. If an authentication system is not installed, the AP can be able to easily get a valid IP address through DHCP. It is assumed in some of these networks that anyone with physical access to a network outlet is authorized to connect to the network. However, the wayward AP will extend the network beyond the reaches of the office. Anyone with a WLAN-compatible computer can connect through the AP without much difficulty. This is like locking the front door, but leaving the back door open.

Using the Existing Wired Network

In a large company WLAN installation, the existence of a conventional wired Ethernet network is a great asset. The wired network provides many services of benefit to the WLAN. Among these are authentication, security, connecting to the Internet, and interconnecting the APs. Properly managed, the wireless network can be melded into a cohesive and versatile resource for the company.

The wired network must have connectivity for all the APs, so that the WLAN users will have the ability to connect to corporate resources and the Internet. Wireless network planning must include provisions for wired network connections. The availability of a wired network connection is one of the critical aspects of determining appropriate AP placement. Unless you intend for the WLAN to service wireless nodes exclusively, you must have a network connection for the AP to bridge the wireless and wired networks together. If you want to have a WLAN that does not connect to a wired network, take a look at the Tech Tip on Wireless-Only LANs.

Tech Tip: Wireless-Only LANs

A temporary network is a very good reason to create a wireless-only LAN. For example, if you wanted to set up registration tables and judges' stations at an athletic event or a temporary network for a meeting or convention, you could use a wireless-only LAN.

The WLAN standard, IEEE 802.11, provides for a pilotless network between individual nodes, such as two or more laptop computers. You do not need an AP at all. However, you must reconfigure your W-NIC for this type of network, which is called *ad hoc*. In addition, you will probably have to provide an IP address to each wireless computer manually because there will be no access to a DHCP server.

The other type of wireless network, with an AP to referee things, is called an *infrastructure* network. In a wireless-only LAN, the AP is not connected to a wired network, so the wired/wireless bridging capability does not function. The AP, however, provides a central point that relays the wireless communications between wireless nodes. As such, it gives you several advantages over an ad hoc network. One advantage is that you may be able to mount the AP and/or antenna in a central location that is easy for the wireless nodes to access. On flat ground, an AP effectively doubles the distance over which a pair of wireless nodes can communicate. If the two nodes are on opposite sides of the AP, you can add the absolute ranges of each link to achieve twice the range.

In addition, you can place the AP antenna at a height that allows communication over obstacles. The AP becomes a repeater for the two nodes. If you have more than two WLAN remote nodes, you can use the highly placed antenna on the AP to effectively increase the range of all the wireless nodes.

The major problem that exists in a wireless-only LAN is that you can lose the ability to automatically assign an IP address and your strong authentication capabilities. The IP address assignment problem can be cured by manually assigning an address to each wireless laptop node or providing a DHCP server with a WLAN connection (or use Windows' default subnet of 169.254.x.x.). Likewise, a WLAN-connected server could provide authentication and encryption.

Connecting the Access Points

Configuring multiple APs in a large wireless network presents two important issues. One is the method by which you provide corporate network connectivity to the APs. Do you want to pop them onto the backbone directly, or do you want to create a special isolated network just for the APs? The other is the need for laptop or PDA users to travel between APs without having to log in all over again.

The connectivity issue is primarily a policy decision. The most prudent choice is to isolate the APs into their own network and then use a single protected gateway for access into the corporate backbone. This is illustrated in Figure 8.6. This scheme is a truly arm's-length network isolation because even the Ethernet switches (represented by the segmented network lines) are dedicated. To implement this structure, you need to provide totally separate network wiring for the AP backbone.

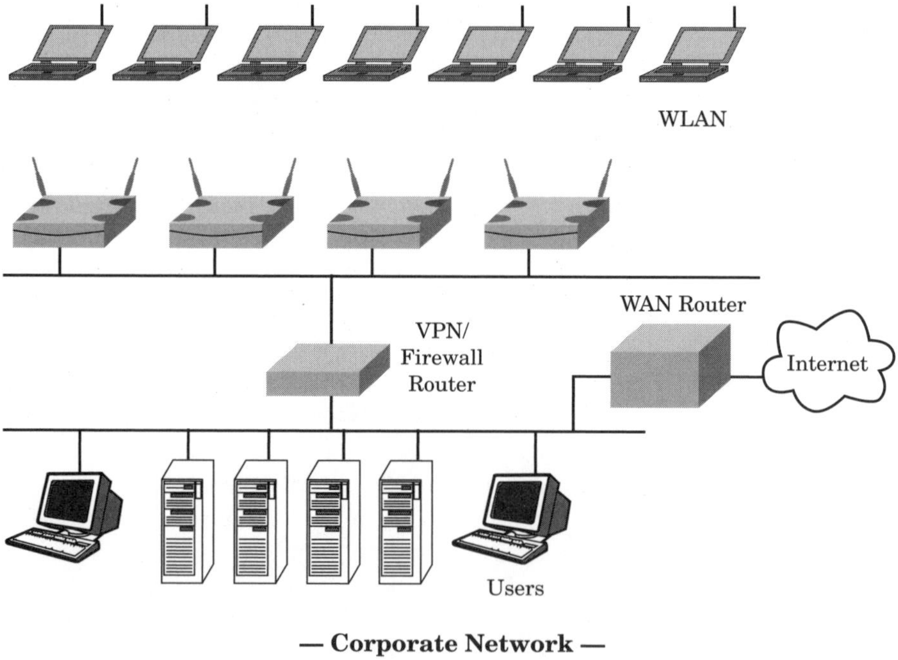

Figure 8.6 Isolating a WLAN from the corporate network.

WLAN

VPN/ Firewall Router

WAN Router

Internet

Users

— **Corporate Network** —

Fortunately, most buildings are wired for networks using *structured wiring*.[4] Structured wiring makes physical separation of the two networks fairly easy. In a structured wiring installation, each network outlet is cabled directly back to a central wiring area, called a *telecommuni-*

cations room. An interconnection to each wiring run is done in the telecommunications room. Thus, a totally separate hub or switch could be provided in that central location, exclusively reserved for wiring runs that go to the APs.

It would be possible to provide an interconnection to the WLAN's VPN/firewall in the server room or in the telecommunications room, if that is more convenient. Because VPN/firewalls are specialized routers, it is also easy to isolate the WLAN and wired network subnets. With the WLAN's APs on the separate LAN backbone, connected to the corporate network only by the VPN/firewall, access to the internal company resources can be tightly regulated, and the possibility of unauthorized wireless access can be blocked.

Using Multiple Access Points

One of the goals of providing a comprehensive WLAN is to allow users the ability to connect to that WLAN from any location that is covered by an AP that is part of the grid. Accomplishing this goal requires a number of factors and also places significant demands on the AP infrastructure software.

To understand how a user connects to an individual AP, we need to explain a little more about WLAN operation in the infrastructure mode. Hang on to your hat because this gets going pretty fast—here goes.

In infrastructure IEEE 802.11 networking, a control node, or an AP, establishes connectivity with one or more wireless stations. The AP allows up to 255 stations to become associated with it and provides coordination between the stations and the AP. The AP and its associated stations form a *basic service set* (BSS). The AP coordinates the transmissions of all of the stations within its BSS, rather like the teacher in the classroom. If every station were to try to talk at the same time, no one could be heard.

APs also can coordinate their activities with other APs to form what is termed a *distribution system*. When the APs are joined together to coordinate their respective BSSs, it is called an *extended service set* (ESS). OK, are you dizzy yet? Fortunately, you don't need to know your BSSs from your ESSs, except to recognize that APs can be provided with special software functions that allow them to send messages to each other, generally over the wired network backbone.

---

[4]Structured wiring is provided by the TIA/EIA 568-B standard, as described in *LAN Wiring*, 2nd ed.

When APs are linked together in an ESS, the software can allow a user (station) to physically move from the coverage area of one AP to that of another AP, without having to reassociate with the second AP. In most instances, that will mean that the user will not have to reauthenticate. This behavior is called *roaming* and is a little different from the concept in cellular telephone networks. In a WLAN, roaming means a station can move freely within the total coverage area of APs that are in the same ESS. This feature is similar to the tower pass-off that cellular radio systems do when you move from one cell to another while making a phone call. A difference with a WLAN station is that its data transmissions tend to be intermittent, so the station and the current AP must communicate periodically to keep the connection open, as it were.

One consequence of creating an ESS is that all of the APs must be able to readily communicate and pass management information and station status to each other. In most cases, this means that all APs must be in the same subnet. This is most important for the roaming station because its IP address is independent of its AP association. Therefore, the AP that a roaming station moves to must support the same IP address subnet, so the station can retain the same IP address, or there must be some mechanism to allow the station to change its IP without losing its connection or network authorization.

AP mutual awareness and linking happens to be a hot area of development right now. As with all complex systems, the IEEE 802.11 standard is undergoing constant refinement. The benefit to the user is that exciting new features eventually will be made available. However, technology always moves faster than standards adoption, so there will always be one or two leading-edge technology innovators who have new proprietary operating modes. Fortunately, it is to the benefit of these innovators that their developments be made operating standards. Of course, they will always have a jump on their competition by being the originator of the new technology.

Planning the Office Network

As we said at the beginning of this chapter, the key to implementing a large wireless network is to do thorough planning. One could also say that a good design is an integral part of the plan. This section leads you through some of the major planning steps. The chapter culminates with a project that goes through an implementation of an office/campus network using these steps.

The Planning Process

To generate a meaningful plan for your office or campus WLAN project, you need to gather all of the raw materials for your plan. For a WLAN, your raw materials are:

1. WLAN coverage requirements
2. Building and site drawings
3. Installation guidelines
4. Manufacturer's specifications
5. Performance guidelines
6. Budget restraints

WLAN coverage requirements. Let's talk about these issues individually. The first item is the coverage requirements for your proposed WLAN. If you are providing a WLAN for a single office space within a multitenant building, your equipment and installation needs will be vastly different than if you must cover an entire multistory building or perhaps several buildings on a campus. (By the way, the term campus has come to mean a group of related commercial buildings in addition to its traditional university meaning.)

Whatever your environment, set out your planned WLAN coverage requirements in writing. Be sure to note any areas that are of little or no value. More than anything, this is to protect you later, when your WLAN will not reach the lobby or the courtyard. If it was not in the plan as a requirement, you cannot be blamed for a lack of coverage there. Of course, you may incidentally provide coverage in areas that are not required. That is a side benefit of the "wandering wireless waves." Sometimes you get some coverage areas for free, whether you want them or not.

Building and site drawings. The next item that you need to obtain is a set of drawings for the buildings your WLAN will cover. If there are multistory buildings, you will need a "plan drawing" for each floor, which shows the overall layout of the floor, including all the office walls and other structures. If you have multiple buildings to cover, be sure to get the site drawings. The building management office should have these drawings available. If you are in leased space, you can probably get the drawings from the leasing agent. In general, building permits and occupancy licenses are required for all commercial construction, so the documents may also be on file with the city or county. Architects are another good source for drawings.

The purpose of obtaining the drawings is to plan where you expect to have WLAN coverage and estimate where APs need to be to provide the desired coverage. The drawings may be copied to the size you need. You normally will not need to keep full-size drawings, but your copies will need to be large enough to be legible. It is a good idea to make several copies, so you can mark up your plans as you go through the design process and the field-measurement verification. Then you will have an extra set to mark the final AP placements and the final coverage areas.

It is best to have a dimensionally accurate drawing because it will allow you to make measurements on paper of the coverage distances. If all else fails, you can make a fairly accurate sketch of the office space based on physical measurements and inspection of the space. Keep track of metal objects, such as structural steel, air ducts, elevator shafts, stairwells, and restroom facilities. These building features may cause coverage problems for your WLAN. Metal screen, railings, and metallic decorations are also potential problems for your WLAN, as are metallic reflective coatings on glass windows.

Installation guidelines. Find out the installation guidelines to which you will have to adhere. This is most important when you install the APs and their antennas. Many ideal AP locations will be along walls or on ceilings where aesthetic considerations prohibit such objects. APs and antennas are not necessarily desirable objects to decorators, and you may have to make accommodations in your design to provide WLAN coverage without offending the eye. If you have written guidelines, agreed to by building management, you will have a defense for the placement of the WLAN components and a valid reason for why complete coverage may not be feasible.

At this phase, you should be able to determine whether APs with certain features are needed for your installation. For example, if the installation guidelines prohibit direct mounting of an AP on a wall or ceiling, you must specify APs that allow external antennas. You also may find that dual W-NIC–APs, which are available for commercial use, have the advantage of covering two areas at once: within the same room and, via a short cable and external antenna, into an adjacent room. Just remember that these two W-NICs should operate on different frequencies because they surely will interfere with each other on the same or nearby frequencies.

This is also a good time to start evaluating the appearance and performance characteristics of external antennas. In most large networks, you will definitely need the flexibility of an external antenna, so you might as well get the options in mind. There is more on antenna systems later in this chapter.

Manufacturer's specifications. Once you have determined your coverage requirements, you can look into the different suppliers of WLAN gear. It is surprising to most people familiar with computer networking that there are wide variations in the performance of WLAN components. This is not to say that some products are inferior by nature, although it is always possible that you might get a bad W-NIC card or AP. However, these devices differ in design and capabilities, so you will have to determine which brands provide the best combination of performance, price, and support for your needs.

Commercial-quality WLAN components are a requirement for a commercial installation. As with anything we buy, some goods are designed for commercial use and some are designed for consumer use. This is just as true of WLAN components (see the Tech Tip on the 6 dBs of Separation). If you get your WLAN APs and W-NICs from a recognized commercial-quality source, you can expect to get more consistent performance, advanced features, expandability, compatibility, and support.

Manufacturer support is a critical issue in a large WLAN network. Designing and installing these systems are not easy. You may want to take advantage of a manufacturer's certified dealer for design assistance and installation. Commercial manufacturers have extensive partnering programs to train and authorize wireless system integrators. Some of these integrators can do the very specialized field-strength measurements that are needed for coverage verification in a complex installation. These wireless system integrators also will have the needed antenna system accessories, such as gain antennas, cables, and amplifiers.

Performance guidelines. Performance guidelines should be determined from the placement of the APs and the building layout drawings. You can determine the expected range from the AP's specifications. Each manufacturer will give its nominal operating ranges for each data rate: 1, 2, 5.5, and 11 Mbps (and possibly 22 to 54 Mbps). These ranges will generally be specified for indoor and outdoor (also called the "clear" range). If you are planning coverage within an open space in a building, such as a large meeting room or an auditorium, you can probably use the outdoor range, if the area is clear of obstructions. If you are in a dense office environment, with lots of obstructions, the indoor range will be more appropriate.

A good way to start is to cut out circular patterns of transparent Mylar™ or acetate. The common plastic sheet protector is fine for this purpose. Determine the radial length of each data-rate range with the scale of the layout drawing you will be using. With a marking pen, mark each range of interest as a set of concentric circles on the plastic sheet and cut out the outer ring of the circle to use for your pattern. Make several of these. Working from the building drawing, position the centers of the circle patterns to simulate AP placement. Use as many circles as you need to cover the entire area, placing the circle center at an appropriate wall or ceiling location. Unless you are working on an open area, you should use the indoor ranges to mark the circles.

Remember that solid walls can attenuate signals considerably. You may wish to neglect outside areas for the purposes of AP placement. Your goal is to gain adequate coverage of the interior building areas, not to necessarily sacrifice interior coverage to prevent outdoor signal leakage. That being said, it is not a bad idea to bias the AP placement toward the inside of the building as long as you can still get good signal strength up to the inside of your building's outside walls.

You will be able to determine how dense to place the APs based on the data-rate rings on your pattern circles. Remember that actual installations may vary considerably due to internal structures and building materials. You should err on the conservative side, if you can. In other words, if you plan for a minimum signal strength that will allow 5.5-Mbps operation in the weaker areas, you will be pleased if some of those areas achieve 11 Mbps and won't be in too much trouble if some dip to 2 Mbps.

Budget restraints. The final planning item is to take your system budget into account. Budget restraints will keep you from overengineering your WLAN. Frankly, having too many closely spaced APs may decrease system performance due to interference. If you have doubts about coverage areas, determine whether you need to use the incremental pay-as-you-go method. When you do this, you provide the minimum number of APs you think you will need, with the proviso that more will be added to bolster WLAN coverage, when necessary.

By all means, you should do field measurements and link testing to confirm your paper design. You may be surprised to find differences from your predictions when you do the actual test simulation. If so, you can adjust your patterns and AP placement to match your measurements. But remember that W-NICs have power and sensitivity variations from unit to unit. So you may need to be conservative when you set up your permanent AP locations.

TECH TIP: THE 6 DBS OF SEPARATION

One of the issues in this chapter is the performance consistency of WLAN components. In one typical manufacturer's AP specification, the power output is stated as 13 to 20 dBm. That's only a 7-dB variation. Gee, that sounds fine—or does it? Let's take a look.

First, power levels of low power transmitters are rated in reference to 1 mW. Then all of the absolute power levels are converted to the logarithmic decibel scale. This conversion uses the formula $10 \log(P_1/P_2)$. Because we are comparing the power to 1 mW, $P_2 = 1$ mW. Thus, a power level of 10 mW would be $10 \log(10/1) = 10$ dBm. Before your eyes glaze over, here is a quick chart that shows several power levels of interest:

1 mW	1 dBm
2 mW	3 dBm
4 mW	6 dBm
10 mW	10 dBm
20 mW	13 dBm
40 mW	16 dBm
100 mW	20 dBm

Now, what was that little 7-dB difference again? From 13 to 20 dBm? If you look at it as being the difference between a puny 20 mW and a robust 100 mW, the difference is significant. That's a 5:1 difference!

What if you had to plan your network with such a variation? The only rational course would be to use the worst-case value, 13 dBm. Is that significant? Well, a 6-dB power difference represents about double the range (or half the range, if you are on the wrong side of the decibel), so you might have to use twice the number of APs to handle the worst case.

Here is another handy chart for dB power levels:

1 dB	1 time
3 dB	2 times
6 dB	4 times
10 dB	10 times

Now you are as prepared as anyone to go forth and design wireless networks. The 6-dB multiple is very useful to keep in mind. It turns out that the basic omnidirectional external antennas offer about 6 dB of gain. The simple Yagi antenna yields about 12 dB, and some of the basic dish reflectors have about 18 dB. We will use these values in calculating our path losses in the Appendix.

Covering the Ground

Up to this point, we have been concerned with creating internal building WLANs. If you want to cover an outside area, your design will be a little different. Typically, outside areas are unobstructed, but developed areas may have large trees and variations in terrain that cause special problems.

As with the building interior, you should start with a layout drawing. The best source for this will be your site manager or possibly the plant drawings that were used for zoning. Architects and landscapers also frequently have site plan drawings that show buildings, terrain, major trees, and other obstructions.

You will need to provide a location that allows your AP to be protected from the weather elements. APs are designed for dry locations and moderate temperatures. To cover an outside area, you will need to mount the AP inside the building, or perhaps in a weatherproof enclosure and run a coaxial cable to the antenna. Be sure to choose an AP with external antenna capability.

You should remember that outside antennas will need solid, wind- and weather-resistant mounts, lightning protection, and the proper line-of-sight access to the areas you wish to cover with the WLAN (Figure 8.7). You may need to use a type of directional antenna that gives a 180° coverage to provide signal to the side area of a building. Or you may need an omnidirectional antenna, mounted on a high pole, to cover a wide area of 360° around the location. Antenna systems are discussed in the next section of this chapter.

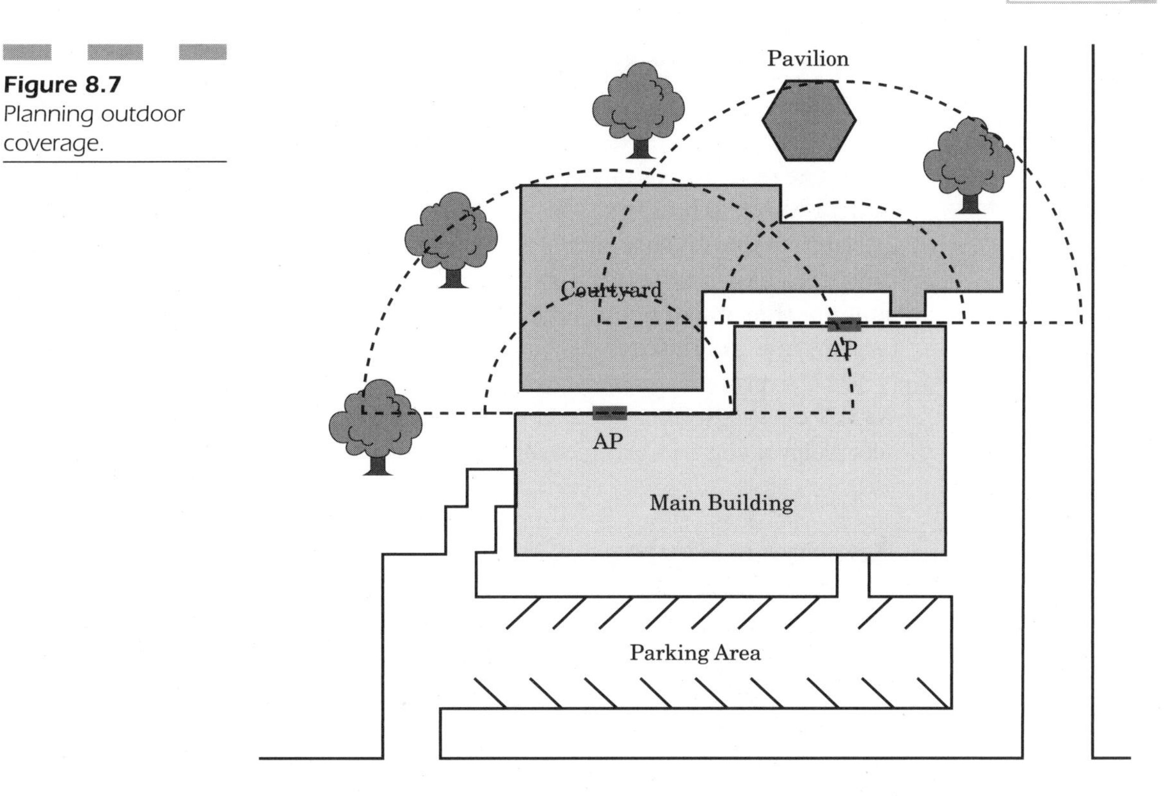

Figure 8.7 Planning outdoor coverage.

You can use the same type of plastic circle patterns that you did for the building interior, only this time use the outdoor ranges and the scale of the site drawing. If you are using a directional gain antenna, modify the markings on the plastic to mimic the antenna pattern and adjust the range circles (or arcs) for the expected range increase.

To confirm your design estimates, make field measurements in the actual outside coverage area. Try to simulate the antenna styles and coax losses of the actual installation. If you cannot get the range and coverage you require, consider adding amplifiers to the system, if permitted. Signal amplifiers can compensate for lengthy coax cable runs and give you much better system performance. In the United States, an output level of 1000 mW is permitted, so you could add an amplifier with a 10- or 15-dB gain to boost your output of a standard 13 dBm AP to compensate for your coax loss and stay well within the limits.

Antenna Systems

The most significant factor that influences WLAN coverage is the performance of the antenna system. We say "system" because much more is critical to the strength of the wireless signals than the individual antenna component. In this section, we cover the basic types of antennas and compare their characteristics. In addition, we talk about the other components that make up an antenna system: cables, mounts, lightning protection, and amplifiers.

With a properly configured antenna system, you can maximize your WLAN coverage from your APs. In addition, with simple antenna system accessories, you can bring coverage to a station in a poor coverage area.

Antenna Styles

Practical antennas can be divided into two primary categories: directional and omnidirectional. In addition, there is a third, theoretical, category, called a *point source*. It is useful to consider this theoretical antenna first because the comparisons of all other types of antennas are based on the mathematical performance of the point source.

A point source is a perfect antenna. The concept presumes that, if you constructed an antenna that could couple energy into space from a single point, it would propagate EM energy equally in all directions. We have a name for the imaginary embodiment of the point source. It is formally called an *isotropic antenna*. This theoretical isotropic antenna is what all other antennas are measured against. It is considered to have unity gain, so the directional gain of any other antenna may be expressed in decibels relative to the isotropic antenna (dBi).

Now back to the two categories of real antennas. An *omnidirectional antenna* is considered to be one that propagates radio waves (or EM energy) equally in all directions *within a plane*. You will recall from geometry that a plane is an infinite flat surface.[5] An example of a typical omnidirectional antenna is the vertical antenna on many cars. Another example is the small wire antenna you see on police and public service vehicles. A third example, is the pull-up antenna on some cellular phones or that little antenna that folds out on some WLAN cards and APs.

---

[5]This is a surface defined by two lines that intersect at a single point.

This type of omnidirectional antenna is intended to function equally in all horizontal directions, that is, all directions radially outward from side of the antenna. In other words, the antenna is oriented orthogonally from its plane of radiation. An example of an omnidirectional antenna is shown in Figure 8.8.

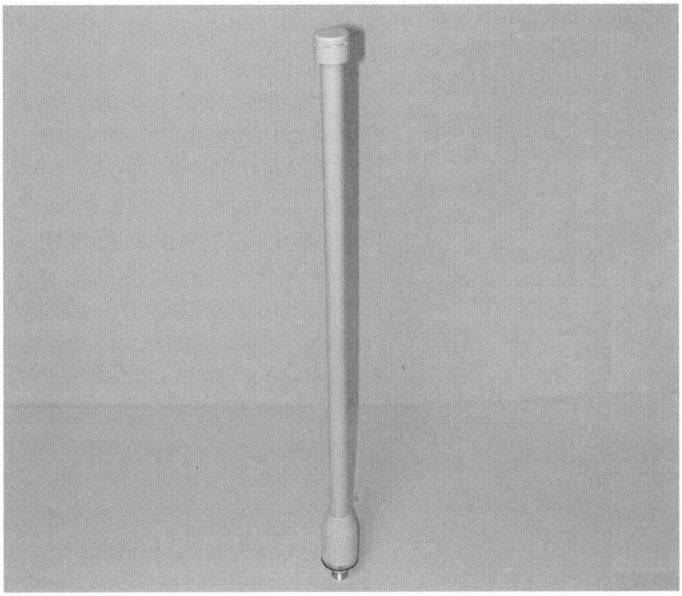

Figure 8.8
An omni-directional antenna.

Omnidirectional antennas radiate[6] some of their energy above and below the plane of orientation. Different versions of omnidirectional antennas have slightly different radiation patterns in the horizontal plane. Some have a fairly broad pattern in the vertical direction, and some have a relatively flat pattern. This is shown in Figure 8.9.

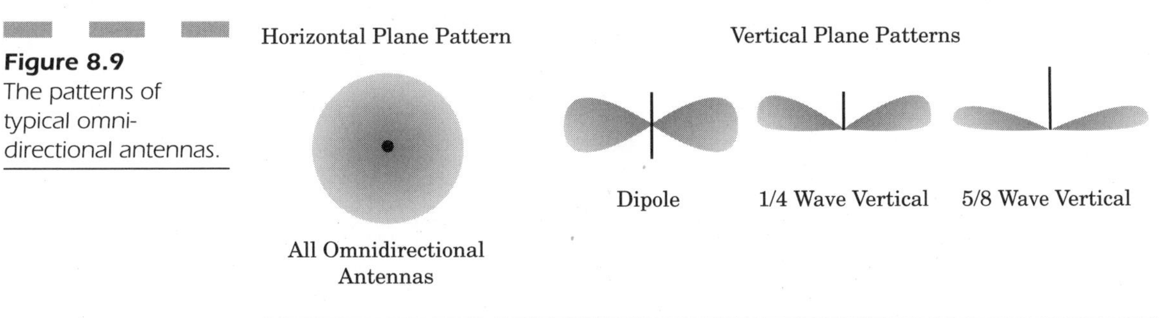

Figure 8.9
The patterns of typical omni-directional antennas.

[6]Radiate refers to both the emission and reception of electromagnetic energy.

The other basic antenna category is called a *directional antenna*. Simply stated, a directional antenna is one that concentrates its radiated energy primarily in one direction. From a practical standpoint, a directional antenna's pattern ranges from very broad to very narrow and concentrated. Directional antennas present the widest variety of styles. Some of the directional antenna styles in common use with WLANs include the Yagi[7] and the parabolic antennas. Examples of these types of antennas are shown in Figure 8.10.

Figure 8.10 Directional antennas (the Yagi and the parabolic).

A Yagi antenna is constructed from a driven element (typically a simple dipole), a reflector, and array of one or more directors. These additional elements are called *parasitic* elements because they steal some of the energy from the driven element. Conveniently, they also reradiate that energy in such a way as to concentrate in the direction of the director elements. Yagi antennas for WLAN use are frequently encased in a plastic housing, resembling a cylinder. The plastic enclosure is sometimes called a *radome*, after the name given the dome-like coverings for radar antennas. This housing is transparent to the radio waves and presents a more appealing antenna assembly. It also physically protects the antenna from the effects of wind and moisture.

[7]The Yagi-Uda array antenna.

This WLAN Yagi antenna uses 14 directors and produces approximately 12 dBi of directional gain. This is about average for Yagis in this frequency range. "Stacking," or using multiple antennas mounted in the same orientation, may add gain.

The parabolic antenna is constructed as an end section of a paraboloid. From geometry, you might recall that a parabola has a focal point. If an EM radiator, such as s simple dipole, is placed at the focus of a metallic parabolic surface, the radio waves will be reflected more or less in the same direction. The effect is to produce a highly collimated beam of RF energy.

The parabolic antenna is very useful for concentrating the coverage pattern of an antenna to a very narrow area. This concentration produces an effective gain of 18 to 25 dBi for the smaller "dish" antennas used with WLANs. For even more gain, larger parabolic surfaces produce proportionally more gain. It is possible to get as much as 50 to 60 dBi gain from a parabolic dish, but such an antenna would have a much larger parabolic surface and probably be undesirable. In addition, the signal feeds and geometry of these very high gain antennas are critical, so these antennas are much more expensive.

Beamwidths

Practical directional antennas have a characteristic pattern that indicates the degree of concentration in a particular direction. We can look at the direction of greatest energy concentration as the zero axis of the antenna. Field measurements will reveal that a directional antenna gradually decreases the radiated field strength as you measure to the left and right of this zero axis. Eventually, the amount of field strength becomes half of the peak strength at the zero axis. These points, in degrees, are called the *half-power points* of the antenna. The radial arc between the half-power points is called the *beamwidth* of the antenna.

The beamwidth of a directional antenna can range from as much as 270° to as little as 0.5°. However, the most common antennas in WLAN use have moderate beamwidths of 5 to 45°. Most many-element Yagi antennas have beamwidths of 15 to 20°. This relatively broad beamwidth allows you to get the greatest benefit from its directional gain without critically aiming the antennas. As long as you get the antenna in approximately the right direction, you are fine.

A parabolic dish antenna for WLAN use will have beamwidths of 5 to 10°. This means that aiming these antennas will be a little more difficult. In general, the higher the gain, the narrower the beamwidth will be. Figure 8.11 shows the typical beamwidth patterns of Yagi and parabolic antennas.

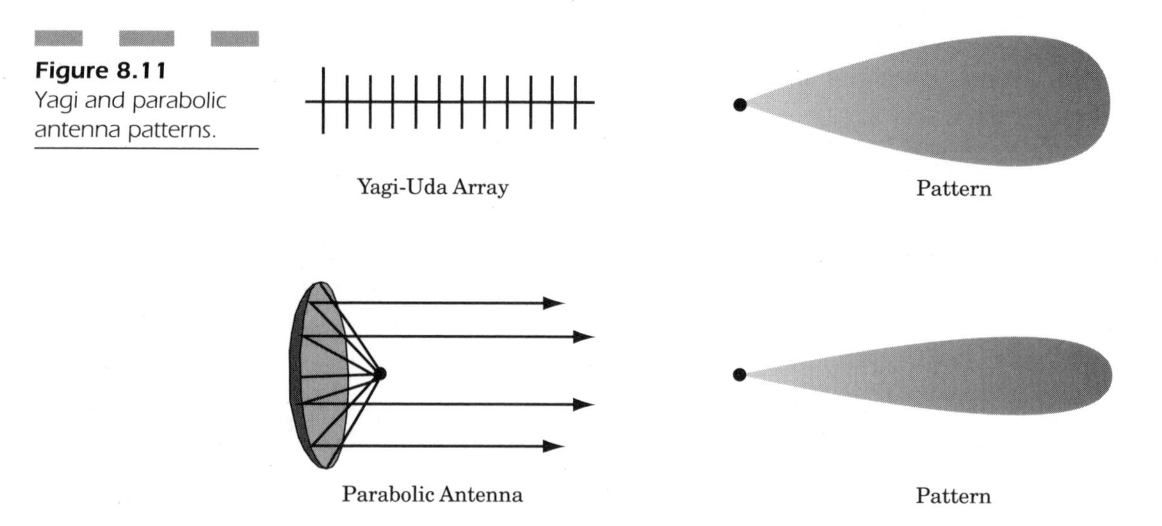

Figure 8.11 Yagi and parabolic antenna patterns.

When looking at the patterns of antennas illustrated in this manner, remember that these lobes represent only the 3-dB points of the antenna's radiation pattern. A true 360° pattern would show a much more gradual drop-off in signal strength and often some signal in virtually all directions. As a matter of fact, most directional antennas have significant *side lobes* that are off axis, where the signal level comes back up. Ironically, the parabolic antenna often has back lobes that are totally behind the reflective dish.

This means that you may need to consider these other directions as having sufficient signal strength to be of use for your coverage or, possibly, a hassle. There is nothing worse than inadvertently creating an interference source where you thought directional antennas would avoid the problem.

Note that Yagi and parabolic antennas use a dipole driven element that gives the antenna a polarization. Be sure to orient the antenna in the vertical position for most applications. Most antennas that are covered by radomes have an arrow or the word *top* to indicate proper orientation. Misaligning these antennas will cause as much as a 20-dB signal loss.

Gain versus Range Increase

The directional gain from antennas is used to help us predict the range of our WLANs. Antenna directivity gain is a significant benefit for the WLAN designer. Essentially, it lets you increase range and extend data rates without using additional APs.

With a little math, we can calculate the effect of distance in reducing WLAN signal strength. We can also easily determine the effective loss from increasing a W-NIC's distance from the AP antenna. Although the math is more effort, you can easily estimate what will happen as your distance increases. The rule of thumb is that the received signal decreases by about 6 dB as the distance doubles.[8]

You can effectively use gain antennas to get this signal loss back. You can also extend the range of a WLAN node and extend the data-rate bands. In addition, you can use antenna gain to compensate for coaxial cable loss, if you have to use a long cable to connect the antenna to the AP.

Lightning Protection

It would not be proper to end a discussion of antennas without mentioning lightning protection. This is probably the most overlooked aspect of outside antenna installation. Lightning discharge is a very serious problem around the world. Some locations are particularly vulnerable to these violent electrical discharges. Even when lightning misses your antenna, the surrounding electrical charges can cause a significant current flow and voltage buildup.

If you mount an antenna on any outside location where it is not well protected, you should provide a proper lightning arrestor and discharge wire. The local and national electrical codes provide for proper lightning protection for receiving and transmitting antennas.

Failure to properly protect your outside WLAN antennas could result in damage to the AP and in extreme cases, to other devices connected to the AP. From a practical standpoint, it is unlikely that your WLAN antenna would actually encourage or direct a lightning strike because most outside WLAN antennas are mounted unobtrusively on the side of buildings or under roof eaves. However, you may need to mount a

[8]For the mathematically curious, the power density at a point of distance R from an isotropic radiator of power P_t is P is proportional to $P_t/4R^2$, and the path loss increases as the square of the distance (and the square of the frequency).

WLAN antenna high on a pole or on the peak of a building roof. In such cases, your antenna may cause a strike that would have landed elsewhere or not at all.

Proper lightning arrestors can help discharge minor strikes and prevent damage to your equipment. However, it is unlikely that a serious strike can be dissipated by normal means. If you have concerns in this area, consult a qualified radio antenna tower technician. You can also contact a company that specializes in providing lightning protection and prevention.

Mountings

A variety of antenna mounts are available to help you properly secure and aim directional antennas. Typical roof mounts include pole mounts, metal straps, and free-standing tripod mounts. The tripod mount allows you to place an antenna on a central pole, stabilized by three legs to the roof. These work best on flat roofs.

Your antenna may come with its own mounting bracket. Directional antennas must be properly aimed to function correctly. You should consider the antenna mounting to be an important part of choosing and installing an external antenna.

Barriers to Wireless Success

The most significant barriers to a successful WLAN installation are interference and physical constraints. Interfering sources abound, particularly with the free and unlicensed use of the 2.4- and 5-GHz bands. It would be ideal if these bands were allocated worldwide for the exclusive use of compatible WLAN technology. However, the intent of providing these frequency ranges was to foster technical innovation, and that is what has happened.

In addition to our use of the two bands for IEEE 802.11 WLAN communication, they are available to anyone who complies with the frequency and power guidelines. Many other technologies have been developed to use these same frequencies. For example, BlueTooth, HomeRF, HyperLAN, and cordless phones use these same bands. In addition, medical, industrial, scientific, and even amateur radio users share these bands. Some interference has even been reported from microwave ovens.

It would be to your benefit to coordinate the activities of those within your WLAN coverage area, so that interference is minimized. At the current time, there is very little peaceful coexistence between the competing technologies. Most are aware of another noncompatible user in the band only when interference occurs. Unfortunately, because these are data transmissions, it may be difficult to distinguish interference from other problems. Some simple interference measuring equipment can be used to check for and troubleshoot interference problems.

Fortunately, efforts are underway to coordinate the operation of several of these 2.4- and 5.2-GHz technologies. IEEE 802.11 and BlueTooth networks are moving along the path of mutual awareness and avoidance. Eventually, it is hoped that all technologies that use these bands will allow compatible operation and that many of the causes of undesirable interference will be eliminated.

Another set of WLAN problems is due to the physical characteristics of the coverage areas at your site. We have tried to cover the most obvious of these in the previous sections on planning your installation. However, some things may change after the initial installation.

For example, in one installation, coverage between two buildings was needed. The two buildings were very close, but the WLAN was deemed to be more cost effective than the alternatives. A WLAN coverage area was set up using a directional antenna, which was connected through an outside wall to the AP. PCs in the other building were equipped with simple signal-booster antennas. Initially, the link operated well, achieving the full 11 Mbps that was expected. However, after more than a year, problems began to occur. The data rate had fallen off and the link failed periodically. The problem was simply Mother Nature. A small tree in the courtyard between the buildings had grown dramatically during the time since the original installation. The dense evergreen leaves were attenuating the WLAN signal. Installing an amplifier and directing the antenna more toward the other building cured the problem.

Schools and Colleges

Schools and colleges present interesting challenges for WLANs. Undoubtedly, this environment offers the most utility and purpose for wireless connectivity. However, many campuses are vast and complex. For example, the typical university may have literally hundreds of buildings spread out across hundreds of acres. Well-developed computer networks and backbones already exist, with their own complement of IP subnets, routers, bridges, and firewalls. To provide a cohesive WLAN fabric across the entire campus requires careful study and analysis. But then, isn't that what universities are for?

Seriously, some of the issues that occur are compounded by the relatively primitive state of AP interconnectivity. A user can encounter a potential problem when moving between the coverage areas of several APs. If the station must maintain its session or its IP address in a large campus environment, each AP may be served through a separate IP subnet. The way IP works, you must have a valid address for the subnet you are connected to, but it would be quite impractical to place all of the campus buildings on a single subnet. For one thing, each campus typically has thousands of IP addresses, and they simply cannot easily coexist in the same subnet. In addition, transport, routing, and access control demand a highly organized and interconnected subnet structure.

Nevertheless, educational organizations, from the largest university to the tiniest public school campus, have an opportunity to provide unlimited connectivity to students, faculty, and staff through the promise of WLAN technology.

Wireless Campus Dorms

A very innovative approach can use WLANs to add Internet access and campus connectivity to student dormitories. Many student housing buildings on campuses were constructed before the age of computers. Some, it seems, were constructed almost before the age of electricity. These dorms require extensive rewiring to provide high-speed computer networking. For buildings that do not lend themselves to conventional data cabling, wireless networking can offer an important option.

Many schools have found that a WLAN can be designed, installed, and operational very quickly. As a matter of fact, for many residence halls, all that is needed is to locate a couple of APs on each floor and provide each with a data connection into the campus network. WLAN cards are available for almost every type of computer, and installation is very easy. Now that the prices of the WLAN components have dropped dramatically, it may be less expensive to equip every student with access to a WLAN than to do conventional LAN wiring.

One moderate-size university found this solution to be perfect for its circumstance. Several older residence halls were scheduled for replacement within the next two years. However, the students and the administration were demanding network connectivity now. This private school faced a daunting task of installing new computer wiring simply to have the buildings torn down to make room for new dorms. They compared the cost of installing a WLAN with installing computer wiring. Because these were older buildings, the conventional wiring was costly. It is always more costly to retrofit LAN wiring than to install it in a new structure.

They found the cost of the WLAN to be within 10 percent of the cost of conventional wiring, even after providing W-NICs for all the students! Moreover, by using the WLAN, they could install everything over a long weekend and have the students up and running the following week. When it comes time to build the new halls, they will probably go ahead and install conventional wiring, but they could just as easily reuse the existing APs for a totally wireless solution. They will also have the APs available to implement WLAN in other campus buildings, such as the library, administration building, and outdoor areas.

Wireless Classrooms, Lecture Halls, and Libraries

The wireless campus is an ideal educational environment. More and more professors are requiring that classwork and papers be submitted electronically. Many campuses have a virtual structure on their Web site that mimics the brick and mortar structure of their departments. Education is greatly enriched by research opportunities available on the Internet.

It seems only natural to bring this environment to the classroom. Many schools are equipping their classrooms and lecture halls with WLANs. This presents terrific opportunities for enhancing the learning experience. Many graduate and undergraduate programs expect their students to have laptop computers loaded with appropriate productivity applications, such as word processing, spreadsheets, e-mail, and Internet browsers. Professors often communicate their assignments, their schedules, and their syllabus online. Students can be directed to real-time resources during class. The computer can be used as an important educational tool, just as the slide rule and calculator were in times past.

Laptop computers can be used in science classes to instantly record experimental results, to do complex calculations, and to perform simulations. They can even be equipped with external peripherals that connect directly to scientific apparatus, to enhance or even direct experiments.

Large amounts of data in a variety of forms can be stored on the portable computers available to the student anytime and anywhere.

Ready availability of laptop or notebook computers also may have some drawbacks, particularly when coupled with WLAN access. More than a few professors have had to ban laptop use in the lecture hall. It seems that the chatter of keys from a number of laptops is quite distracting to other students and the professor. However, it would be great to complete the homework assignment, catch up on your e-mail, or surf the Web during one of those particularly boring lectures.

Project: A Large Office/Campus WLAN

The project activity for this chapter involves creating a WLAN for a large office or a commercial (or educational) campus. We do not pretend that you can design an actual office/campus WLAN of this magnitude for this chapter's project. Unlike our prior projects, which were fairly general purpose, fixed-scope projects, designing a WLAN system for a large office or a campus of buildings is a custom-designed task. We really cannot know enough about your specific office/campus project to have a normal parts list and a step-by-step procedure.

Instead, we will outline each project step required to complete a typical large-scale WLAN project. The project discussion reiterates much of what we covered in this and previous chapters but presents the information in a linear set of steps to assist your project planning and implementation. You can use these steps as a checklist for your project. We caution you to not leave out any of the steps in this outline. Although you can install a modest office WLAN "by the seat of your pants" using guesstimates and hunches, you will fail if you try to use that approach on a large project. Many of the considerations that were minor points for the home or small office become major stumbling blocks in a large office project. The potential problems do not result solely from the scale of the project but also from the interactions of all the component parts that you must install.

Project Description

In a project of this magnitude, it can be said that "planning makes perfect." The best way to ensure success is to complete all of the planning and design steps before you purchase the first WLAN component. The planning process encompasses all of the activities that precede implementation. Planning includes everything from determining the scope of the WLAN coverage to designing the optimal implementation. A properly conceived large WLAN plan also includes research, field measurements, project budgeting, and management approval. The final stage in the preinstallation process may be to issue a request for quotation (RFQ) or request for proposal (RFP).

Depending on your organization's procedures, it may be possible to have a prospective bidder propose a solution, including a rough design and implementation plan. However, the resulting proposal will be much more complete if you have done a preliminary design and the field test verification before issuing the RFP. This will allow you to control your WLAN coverage and performance much better than leaving the important design details to the lowest bidder. It is only natural for a bidder to favorably lower its bid by skimping on the number of APs and compromising on their installation locations. Obviously, if you have a good preliminary design, you control these issues and consequently ensure that any successful bidder will provide the WLAN coverage that your organization requires.

After a workable plan has been created, the implementation phase begins. We describe several steps that will occur during this part of the project. Installation of your WLAN project may be accomplished by an outside contractor or by your own staff members. Some adjustments to the installation will probably need to be made during and after the course of the installation because of the nature of wireless communication. It is important to include mechanisms for these changes in the planning, approval, design, procurement, and implementation processes. The end result will be a fully functional, efficient, WLAN that totally meets your project requirements.

We have said that you should do all the planning before you buy the first WLAN component. This is not to say that you should avoid buying even a single WLAN starter kit to get familiar with the technology. However, you should not buy and install large numbers of these wireless components without some serious preparation and planning.

Project Step by Step

Let's start with a review of the planning and installation steps that we covered in earlier sections of this chapter. We put these in a procedural order this time, so you can easily visualize the project steps. This allows

Project: A Large Office/Campus WLAN

you to develop a project plan, including milestones, resources (people), and critical paths. Here is a list of the steps that will be covered:

- Obtain site and building drawings
- Outline coverage goals and requirements
- Determine project budget (and phases)
- Create WLAN design and installation guidelines
- Gather manufacturer's information
- Create detailed planning document
- Make field measurements to verify design
- Finalize design
- Publish statement of work
- Bid equipment procurement and installation
- Complete the installation
- Evaluate WLAN coverage and adjust as required

In the step-by-step portion, some of the steps are grouped because they are related. Several of these steps will take some time to complete, and you would be wise to start several of the steps early because it may take some time to complete them. You also want to have several of the project steps running concurrently. For example, it may take some time to obtain all the drawings you need and gather specifications and budgetary pricing from manufacturers. You can also identify potential vendors for the hardware and for the installation, if you do not plan to do this in-house. All these things can be started while you are working through your requirements, the project budget, and other guidelines.

You should lay out your project steps individually, adding as much detail as you can. You can use one of the project planning software tools for this, or you can sketch it out on paper or on a marker board.

Step 1: Obtain Site and Building Drawings

To implement a large-scale project, you should start by gathering all the basic information you will need. You can obtain drawings from architects, building managers, real estate agents, and even city planning departments. If you have a single building (or one or more floors in a building), you will need the building layout drawings for the floors you will be covering with your WLAN.

Project: A Large Office/Campus WLAN

For multiple-building projects, you will also need an accurate site drawing. It may be necessary to create a version of the site drawing that shows building heights, so you can figure line-of-sight coverage areas. Any building may shield WLAN signals to a certain extent, but a taller, larger building will exacerbate the problem. You will also need to know suitable locations for outside antennas, if your plan calls for them.

This is also the time to check for building code and covenant restrictions. Some jurisdictions have a permitting process for virtually all construction, and some of the minor building modifications required for WLAN implementation may fall under these guidelines. In addition, you will need to meet all local and national electrical codes for the wired portion of the WLAN network, such as the wiring for the APs and lightning protection for any outside antennas. Some commercial property is subject to private regulations, called restrictive covenants. This may limit your ability to use outside antennas, such as parabolic dishes.

You might also need to get permits for antenna structures on tall buildings. Certain structures in an airport flight path must obtain authorization from the national aviation authority (the Federal Aviation Agency in the United States). In some circumstances, special lighting or strobe lights may be needed. If the building or tower height was previously approved, you may need an amended authorization if you add an antenna that increases the overall height.

Step 2: Outline Coverage Goals and Requirements

Step 3: Determine Project Budget (and Phases)

Outlining the coverage goals, discussing the budget, identifying special installation issues, and refining the design are part of the planning process. For these steps, you will need to involve people from different departments. It is best to have some general idea of the capabilities of a WLAN, so you can set expectations realistically. An important part of this process is differentiating between mandatory coverage locations and optional coverage locations. Determining the performance levels, in terms of speed, at each mandatory location is also crucial.

You might want to set up three categories of coverage: mandatory, adequate, and desirable (or, you could categorize these coverage areas as A, B, and C). Mandatory coverage describes an area that absolutely must have top-level WLAN coverage, such as a conference room, auditorium, or the executive offices. Adequate coverage areas would be those that would have coverage but might not have top performance. Desirable coverage concerns the optional areas, where you really would not mind having coverage, but management doesn't want to spend the money to guarantee it, or where a minimum performance level is acceptable. You should also determine the performance levels (i.e., speeds and signal strengths) you expect for each area. Keep management informed that absolute speed depends on factors such as signal strength (which is related to AP location and antenna gain) and environmental noise.

You should expect to present a formal budget for the project. You can start with a preliminary budget, based on your best estimates of the costs of the WLAN components and installation. You naturally will need to have assumptions about your expected coverage goals before you can sum the wireless infrastructure costs. You can get some approximate pricing guidelines from WLAN equipment suppliers who will also be able to project pro rata costs. For example, you should be able to get ballpark costs for each AP and W-NIC and an average cost of installing an AP. Be sure to get estimates for providing AC power and network connections. You should anticipate that some APs will need external cables and antennas.

The budget at this point should be very flexible, unless you are working from a fixed grant or fund allocation, which reverses the budget process. Starting with your desired WLAN coverage, you can add up the projected project costs including a factor for postinstallation coverage adjustments. A good rule of thumb is to allow 15 percent for changes necessary to bring coverage up to expectations. Then your preliminary budget can be brought forward for discussion. At this stage, you will not have a final design or bids, so you should leave the final budget figure open. This process may proceed at the same time as steps 4–7, making adjustments as needed.

Step 4: Create WLAN Design and Installation Guidelines

Now you can begin to design your WLAN implementation. You will need to locate all the APs and figure out how to get power and network connections to them. As this process continues, you will construct your detailed list of component parts and set the specifications for installation. The design will identify any special considerations, such as directional gain antennas, towers, wireless bridges, and WLAN channel plan. Include AP configuration and remote management as part of the installation.

Project: A Large Office/Campus WLAN

The best approach is to set management's expectations to match the worst-case performance of the WLAN installation. Nobody will be upset if the system works *better* than expected! Wireless technology is capable of certain levels of range, speed, security, and reliability. If you set the expectations so that the users and the managers realize that limitations exist, they will be much happier and your complaints will be minimal. Management should also know that coverage can be added to strengthen weak areas, and that a certain amount of adjustment to the original design is normal after the system is installed.

As you gather your coverage and performance goals, begin to mark them on your drawings (keeping unmarked copies for later use). Proceed one building at a time, floor by floor. As you complete the coverage areas in each building, begin to plan the outside areas that need coverage. Next, determine how you will link WLAN coverage areas that are in separate buildings. Will you need to create a WLAN point-to-point link, or do you have fiber optic cable between the buildings?

You should begin to set the policy for linking the wireless and wired LANs together. You can connect the APs in each building to that building's subnet, if you organize your IP space that way. If the buildings are bridged, it is possible for them to use the same subnet. However, if they are connected with routers (which is common) your APs will be on a separate subnet in each building. That is a problem only for WLAN roaming (going between APs without relogging in), without special networking software.

You should also establish the authentication and security policies. This might affect the manufacturers you choose for your APs and W-NICs because they will have to support the encryption and authentication systems you choose. If you have an existing authentication, authorization, and access server (sometimes called AAA or Triple A), you can probably choose WLAN equipment that will be compatible with what you are using. In a large private company, you certainly will be concerned with security, and this subject should be covered thoroughly with your security administrator. Many companies have a chief security officer who deals with these policies.

Public organizations, such as government and education, have similar security concerns and policies. Colleges and universities often must manage dual systems with loose security rules for routine student access and very strict security rules for administrative and research access. It is not unusual for these organizations to maintain multiple levels of subnets, virtual LANs, access control lists, and VPNs to provide a complex structure for access to networked resources.

Step 5: Gather Manufacturers' Information

This step can be started at almost any point and continued in parallel with the other steps. Manufacturers' data sheets will contain the specifications that you can use to set your design parameters. You can also get pricing and availability for the brands that you are considering. At this time, you will identify potential suppliers and installers for your WLAN system. All of these third parties will add to the information you need to make a comprehensive project plan.

Additional accessories and capabilities can be identified at this stage, so you will know where to go for these resources when you need them. This is similar to the process that all designers and architects go through, but it is specific to a WLAN. For example, are there any significant performance differences between manufacturers? What special antennas, preassembled cables, mounts, and amplifiers are available, and what do these items cost? As you continue your design and continue refining your budget, you will have the information you need to make trade-offs to accomplish your goals.

If you have any concerns about performance or compatibility, this is a good time to bring in evaluation equipment for testing. You can do some simulated range testing and also determine whether all the proposed types of equipment are compatible with your needs. For example, you may want to use one brand of AP for most of your installations but another brand of AP to accommodate dual W-NICs in a few locations. It is important to check that both units can peacefully coexist and support all of the features you need (such as roaming).

You also might want to check out the operation of proposed APs with the variety of W-NICs that you will be using. Be sure to verify that all wireless devices will support the authentication, encryption, and OSs you expect to use. This is also a good time to check out each manufacturer's support policies. Technical support and product warranty procedures are critical to a large organization, and you will want to make sure that prospective vendors offer adequate levels to meet your needs.

Project: A Large Office/Campus WLAN

Step 6: Create Detailed Planning Document

The detailed planning document will include the placement of APs, the provision for power, the locations of antennas, and the connections to the wired network backbone. If you use computerized drawings, it would be good to create one overlay for your WLAN coverage plans and another for the AP and antenna locations. The AP drawing layer should show the expected range for the target WLAN speed(s). Another layer could include the power and network connections. Include the proposed IP addresses on another drawing layer.

This planning document incorporates everything you have worked on to this point and ultimately will be modified to become the final plan for your large-scale WLAN system. The planning document includes the project's coverage goals, the budget, the design and layout, and the installation guidelines. It outlines the requirements and sets performance expectations. The document will be adjusted, if needed, after your field measurements (step 7).

Step 7: Make Field Measurements to Verify Design

After you have created your detailed planning document, you need to make field measurements to verify the design. This step is crucial to verify that your initial design provides the desired levels of WLAN coverage. Field measurements are particularly important in the mandatory coverage areas. You will learn where coverage holes are and will be able to test alternate AP locations to boost coverage. With the information that field measurements provide, you will be able to make adjustments to the number and locations of the APs and to the gain and location of any antennas. This will allow you to create your final design.

Review the information on field measurements that were covered earlier in this chapter. At a minimum, you should simulate AP installation and use a calibrated W-NIC receiver to gauge signal strengths in the proposed coverage areas. However, the best quality (and most expensive) results come from formal field-strength measurements, usually available only from a third-party WLAN testing service. The field measurement provider also can make recommendations about AP placement and help you with the choice of antennas and other accessories.

Project: A Large Office/Campus WLAN

Step 8: Finalize Design

Step 9: Publish Statement of Work

You can make any adjustments for the design that come as a result of the field measurements. The final design is then incorporated into the final project plan, the proposed budget adjusted, and a statement of work is created. This statement of work, which references the final WLAN design and installation guidelines, will be the guideline for installation. In conjunction with the list of WLAN components, the statement of work is crucial to the contracting process, if the installation is to be done by a third party.

If you are doing the project with internal personnel, you can use a less formal outline for the statement of work and the installation guidelines. You will still need to be able to define and monitor the progress of the project, and a fairly formal process will be the best way to accomplish this. You also get the benefit of management approval and acknowledgment of the "fine tuning" that may be needed to get WLAN coverage in all desired areas.

Step 10: Bid Equipment Procurement and Installation

At this stage, you will be able to estimate the cost of the project and compare it with your budget constraints. You may also get formal bids or quotations for the project. Most large projects will be professionally installed. It is always best to get your WLAN components and installation from the same company. It does not matter whether the WLAN equipment vendor provides installation as a value-added service or a qualified WLAN installer provides equipment. However, at a minimum, the installer should be trained and qualified to install the brands of WLAN components you are using.

When you get bids for your project, remember that absolute coverage guarantees are very difficult to honor. This is much different from conventional network wiring, where there either *is* or *is not* a connection, and it generally works perfectly, if it works at all. In a WLAN installation, factors such as unexpected interference or shielding may cause problems. In addition, the WLAN vendor may not have any control over the user's W-NIC brand, its relative performance level, or how it is installed. As you have seen in the earlier chapters, relatively minor factors, such as laptop PC orientation or metal cabinetry, can have a detrimental effect on a wireless link. A much better approach is to require a certain level of signal (field strength or received signal level on a calibrated W-NIC) at discrete points in each coverage zone. This approach is similar to the verification method used in a wired network, and can be easily verified by measurement, if there is a concern that the installation did not meet specifications. Often minor problems can be cured with a different type of antenna or a booster amplifier.

Certain portions of a WLAN installation may require very specialized capabilities, and a subsidiary vendor is definitely acceptable. For example, a properly trained and supervised third party may provide installation of antennas and WLAN equipment in outside locations, such as outer walls or antenna towers.

Step 11: Complete the Installation

The installation of a complex WLAN project will be multifaceted. On one level, you will need to install the wireless infrastructure for your WLAN system. This will involve providing the AP equipment and network connectivity for the wired backbone. The network cabling will have to be extended to each AP, and the method of providing roaming (if it is needed) will have to be set up. Quite a bit of setup will be needed for each AP in a complex network. You will need to change all the defaults, set up the channel plan, and provide remote administration.

In addition to providing the obvious wireless medium, you will need to set up the environment on your wired backbone and its interconnection to your corporate/enterprise network and, ultimately, to the Internet. This will involve the implementation of the chosen subnet addressing scheme, authentication, encryption, firewalling, and routing needed to support the level of WLAN access you have proposed.

Step 12: Evaluate WLAN Coverage and Adjust as Required

As you install the WLAN system, you can begin to evaluate the system performance. You may encounter unanticipated problems. If so, you can use your planning documents to devise a workable solution. Feasible alternatives exist in almost every case. To overcome a power or mounting problem, you may need to provide two APs, at slightly additional cost, or you may need to substitute a dual-NIC AP. Fortunately, the project plan has allowed for a slight overrun in your budget commitments, so you have provisions for most minor contingencies.

Project Conclusion

At the successful completion of the Large Office/Campus WLAN project, you will have implemented an important wireless expansion of your organization's network. You will now have WLAN access throughout your office building or campus. You will be able to provide high-speed WLAN coverage in the areas that were planned for and will have anticipated the complex series of actions needed to ensure a successful WLAN installation.

Traveling Wirelessly

Chapter Highlights

- WLANs in Airports and Hotels
- WLAN Providers
- Improving Coverage Areas
- Security Risks

Personal computers are becoming increasingly more and more necessary. This situation is true for our professional and personal lives. If you are in the business world, you need to keep many of the aspects of the modern office available to you when you are not physically at the office. The office environment, in the Computer Age, includes access to the Internet and the company intranet. Also necessary are access to e-mail, customer relationship systems, and database files.

It is not only possible but also expected that the businessperson have the ability to connect to the company network while on the road. For some of us, that means a place to plug our computer into the Internet at the hotel in the evening. For others, it means having a portable office at all times, even in the airport.

E-mail availability is another critical necessity. It is ironic that e-mail has moved to such prominence. Many of those who send electronic mail expect not only that it will be delivered instantaneously but also that the intended recipient will constantly have access to their own e-mail servers and be able to promptly retrieve the e-mail. For travelers, that presents an entirely new set of problems. You now need to regularly check your e-mail, just as you do your voice mail. Many important messages are sent electronically, out of habit and convenience, with no regard to the recipients' ability to collect those messages in a timely fashion.

You really need to be able to work electronically from anywhere. When you are out of the office, you still need access to the computer and network resources you had at headquarters. This is important, even if headquarters is at home.

This chapter outlines some of the innovative places you can connect via WLAN to the Internet and, through it, to your home or office networks. With the addition of WLAN connectivity to your portable PC, you can truly travel wirelessly.

WLANs in Public Places

The availability of WLAN access in public locations is a tremendous advantage to the individual portable computing user. With a W-NIC in your laptop or palmtop computer, you can quickly and easily connect to the Internet. Once you are on the Internet, you can access all of the practical, useful, and entertaining resources on the World Wide Web and connect remotely to your corporate network.

For example, you can make airline reservations, send Web-based e-mail, trade stocks, or even pay bills. You can even check on your credit card accounts to make sure you can get home. You can look up information on the city you are traveling to or the company you are visiting. You can do research on topics in your area, read the on-line sections of major newspapers and magazines, check the latest national and international news, or check up on a competitor.

With the proper configuration of software, you can create a secure VPN connection back to your company network. Through the VPN, you can do almost everything you could do at the office…except hang around the water cooler. You can perform all the ordinary PC-oriented tasks as well as send and receive faxes, documents, and even voice mail.

WLANs for Travelers

WLANs are being setup in a variety of public and private locations for the use of travelers (Figure 9.1). One of the most logical sites for a public WLAN is in an airport. Frequent flyers are often kept waiting between flights. At one time, a periodic phone call back to the office was all that was expected of the business traveler. Eventually, the ready availability of cellular telephones increased that expectation to *frequent* phone calls.

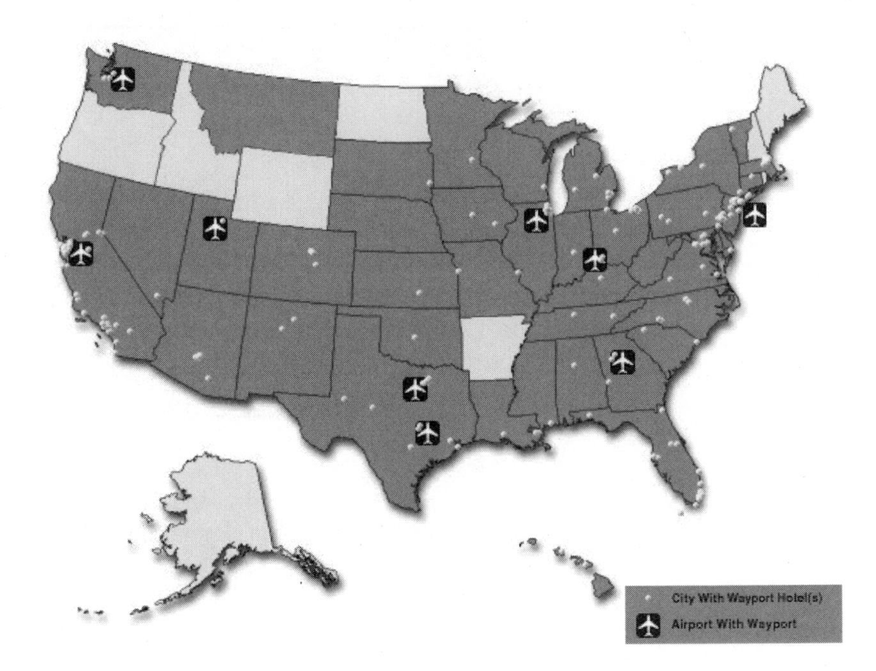

Figure 9.1 WLAN locations are everywhere. Courtesy of Waypoint, Inc.

Now, many office functions, from e-mail to customer relationship management, are required activities even while out of the office. The business traveler is expected to manage a normal volume of these network-centric activities, even from the road. One alternative to working late nights in the hotel on a dial-up line is to use the idle time at the airport to connect into the company network.

Many airports have business centers or temporary workspaces that offer a desk, a telephone, and Internet access for an hourly fee (Figure 9.2). However, most only offer a dial-up telephone line. A few are offering wired Internet access, but there may be added complexities to making a connection. You may have to know how to reconfigure your network settings and change them to match the local LAN. Also, you may or may not get much help or technical support in these facilities. Their personnel are there primarily to receive payment for their office services, and can offer limited assistance connecting to their internal network. Unfortunately, every one of the facilities is set up a little differently, so one configuration may not fit all locations where you want to make an Ethernet connection.

Figure 9.2 *Airport business center. Courtesy of Waypoint, Inc.*

A good solution is being offered at many airports, hotels, and conference centers. The operators of these facilities now offer a WLAN connections to the Internet. This gives them the ability to provide a value-added service to their customers, with very little cost. The travel industry is very interested in WLAN technology because it allows them to differentiate themselves from the competition.

You might wonder where WLAN services are being offered. If you are not already a WLAN user, the answer might surprise you. WLAN access is being offered at airports, airport private clubs, restaurants, and waiting rooms worldwide. Although it is not in every possible location it probably will be. From Seattle to Miami, from Buffalo to Los Angeles, from Dallas to Chicago, airports have WLAN services installed. They are all across North America. And you can find them in Europe, Asia, South America, and Australia. It's a WLAN world!

In addition to the airport locations, you can access WLANs in hotels, motels, resorts, and convention centers. Many luxury hotel chains offer WLAN connectivity as a perk for their customers. Eventually, you may find WiFi™ WLANs featured in advertising and travel guides, the way amenities such as a pool or a color TV are featured now.

There are very practical reasons that hotels would want to deploy WLAN technology. The increasing use of laptop computers and dial-up modem lines by frequent travelers has caused a real problem with their in-house phone systems. Phone systems are typically build for a 5 to 10 percent load. That load factor assumes that, at any one time, only 1/10th to 1/20th of the room phones would be making an outside call. This capacity is probably adequate, as long as their guests only use their room phones for occasional, brief calls. But if a number of those travelers start using their room phone lines for modem connections, the length of time and number of lines in use go up dramatically. In a hotel with a high percentage of business travelers, their load could easily hit 50 percent. To compensate, the hotelier must choose between adding more phone lines and phone system capacity or putting up with very displeased guests.

A viable and cost-effective alternative is to provide an in-house LAN for their guests. In addition to the obvious relief of load on the phone system, the guest gets the benefit of a much higher-speed Internet connection than the phone modem.

A wired LAN is practical in new hotels or ones that can easily be rewired. This is not necessarily the case in existing hotel or motel properties. Unfortunately, older telephone-grade wiring is often not capable of operating a modern LAN, so the hotel owner must be willing to pay for extensive rewiring to add this feature.

Even if the hotel rooms were to be wired for Ethernet, the common areas and the meeting rooms would still be difficult to service. It is difficult and somewhat dangerous to string network wiring all over a meeting room just to allow the participants optional access to a high-speed

LAN. And providing wired access on a case-by-case basis is awkward and expensive.

The WLAN is clearly the answer. The hotel can provide WLAN coverage in all the meeting rooms, common areas, restaurant, and convention center. A hotel visitor can connect into the WLAN with very little fuss (Figure 9.3). A third-party provider normally arranges for the WLAN service for the hotel. The provider installs all the WLAN equipment, including wireless APs, interconnecting cables and hubs, and high-speed Internet access. In addition, the provider normally offers technical assistance and may even sell W-NICs and installation to prospective wireless users at the hotel.

Figure 9.3 A hotel WLAN connects guests. Courtesy of Wayport, Inc.

The powerful concept of the hotel WLAN is that meeting attendees and paying guests can use the wireless connections. At some hotels, the wireless users log in to the WLAN through a hotel Web site that provides WLAN authorization and hotel-specific information. This added service may feature valuable traveler information, such as restaurant recommendations, tour offerings, transportation services, and other guest services. Other hotel services, such as room service, reservations, travel planning, and guest surveys, can be offered.

Another way that WLANs can be used advantageously for hotels is as an alternative to rewiring the guest rooms for Ethernet. In many hotels, adding to or changing the telephone wiring is a tenuous proposition. Few hotel rooms have the drop ceilings that make computer wiring so convenient in business offices. Similarly, few hotels have attic and crawl spaces to make adding computer wiring practical, as often found in private homes. Properly placed APs and antennas can enable the hotel to bring WLAN services to many, if not all, of its rooms. All this is done with a minimum of disruption to the guest rooms.

Remember the WLAN is potentially an 11- to 54-Mbps data connection. Even at the limits of wireless range, which one may encounter rapidly due to absorption from walls and floors, the WLAN still boasts 1 Mbps.[1] This is easily 20 to 50 times faster than a dial-up connection. As any frequent traveler knows, hotel phone lines almost never allow a 56 kbps dial modem to operate at anywhere near that speed. So you are looking at a best case of around 26.6 kbps, dial-up, versus a worst case of 1000 kbps (=1 Mbps) over the WLAN. Guess which everyone would prefer!

All in all, a WLAN for a hotel is a win-win-win situation. The hotel wins by being able to offer valuable services to its guests with a lowered impact on their own telephone systems. The guest wins through the availability of desired high-speed Internet access. The provider wins by generating revenue from the hotel guest.

WLANs for Customers, Patrons, and Students

WLANs are becoming available in more and more places. Some of these innovators are finding that having a WLAN can increase their business and customer satisfaction. Others provide the WLAN as a convenience for their patrons. Still other organizations provide WLAN connectivity as a means of fulfilling their function or of doing business.

In the recent past, the Internet café was born. The basic plan was that customers would come to the venue to surf the Internet. While they were there, the café could serve them food, beverages, and good company. The only real issue had to do with the computers that provided the Internet connection.

There are a variety of options available to the Internet café owner. Some proprietors rent Internet time by the hour (or more likely, by the minute). Others provide the computers at no charge, hoping to attract

[1]Admittedly, the access speed to the Internet is more dependent on the speed of the hotel's high-speed Internet connection, which is provided by the WLAN contractor.

more customers. In some instances, third parties provide the computers and administer the customer access charges.

Many potential customers like to bring their own computers to Internet cafés and similar locations. This is very reasonable, if you want to do more than surf around on the Internet for your own amusement. If you are trying to do some work or return personal electronic correspondence, you are more likely to want to have your own computer handy. Unfortunately, connecting to a wired network in an Internet café is somewhat difficult, if it is even feasible. First, you must deal with the location of the Ethernet jacks. Then you have to bring your own 10/100BaseT cable (which may not be long enough) or borrow one from the management. Finally, you will probably have to change the configuration of some of your computer's network properties to connect to the café's local subnet.

Ultimately, this is just too much of a bother for most restaurants and coffee houses. However, a viable alternative exists through the use of WLAN technology. All the café needs to do is install a simple WLAN AP and interconnect it with high-speed Internet access. Then any customer with a W-NIC–equipped laptop computer can come in, sit down, and surf. This is exactly the same concept as the public access WLAN in an airport or hotel.

Several large coffee shop chains have begun a nationwide (and worldwide) program to equip their locations with WLANs. All a customer needs to do is subscribe to traveling WLAN service through their WLAN provider, and they can use the coffee shop as their personal wireless office (Figure 9.4). Imagine that, surfing and working, while sipping your jumbo latté.

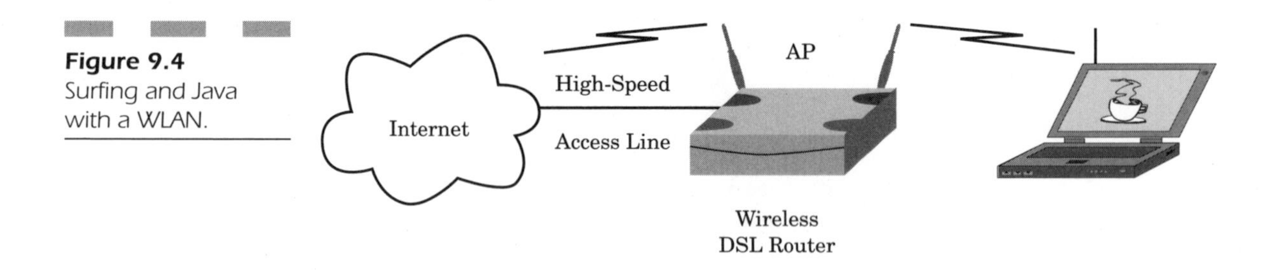

Figure 9.4 Surfing and Java with a WLAN.

Several other types of businesses are adding WLANs as part of their amenities. For example, many bookstores find that customers who can bring in a portable computer and get connected will spend more time and dollars in the store. And many restaurants find that adding wireless Internet access through a WLAN is a great way to increase their business and revenue during their off-hours.

Another type of organization that is rapidly adding WLAN access is the educational community. From private schools to public universities, WLAN access is an important means to meeting educational goals. Every educator will tell you that the resources that are available on the Internet add enormously to the potential educational experience. If students are studying culture, language, or events in another country, they can access awesome resources over the Internet. The Internet has truly become an important research tool and one that can open the student's horizons.

Would you like to find an international studies program that offers a graduate degree and is conducted, in English, in four European countries? It's available in the Internet, with information on a full scholarship. What about a complete searchable text of the Constitution, current copies of the mandatory reports corporations file with the Securities and Exchange Commission, or the complete text of all of Lewis Carroll's works? It is all there, and much, much more.

The opportunity for interaction with people in other countries has expanded. For example, imagine a teen-age student working on a science project, who was able to locate and scan the research of one of the world's foremost experts on animal cellular division. And then imagine that the student was able to receive a favorable review of the project from the renowned expert by using e-mail. Such capabilities add measurably to the educational experience.

To make these and other exciting research experiences available to all students, the school, college, or university must provide a means of Internet access for their students. The typical school, ranging from grade schools to universities, may have 500 to 50,000 students, all of whom present quite a connectivity problem. Many schools are solving this problem through the use of WLANs and high-speed Internet access.

Primary and secondary schools are providing different approaches to Internet and network access. One of the biggest problems for these schools is providing computer labs in fixed locations because of the limited classroom space and scheduling difficulties. Some schools have solved this problem by acquiring a wireless portable computer lab, with a wireless AP node and wireless laptop computers on a motorized cart. The computer cart can be moved from classroom to classroom, as needed, without worrying about wiring or extra classroom space. The AP is simply plugged into a single wired Ethernet jack in the classroom, and the laptop computers are connected to the network wirelessly, using 802.11 W-NIC cards, as illustrated in Figure 9.5.

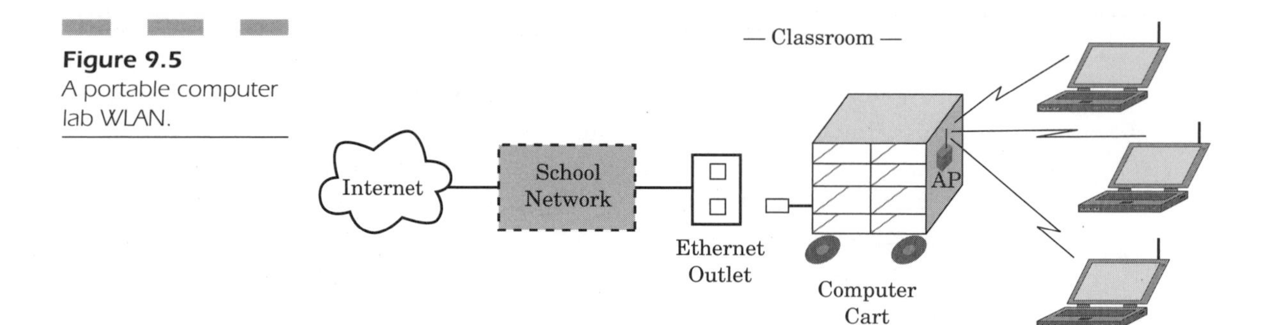

Figure 9.5 A portable computer lab WLAN.

Some colleges and universities are requiring students to obtain a compatible wireless laptop computer for use in completing their courses of study. This approach is particularly prevalent in graduate schools. In some cases, the tuition includes the purchase of the laptop and W-NIC. Some high schools are trying this approach by providing wireless laptop computers to every student. Computers and Internet access are becoming critical to the education process, and it is imperative that all students have access to these resources.

In addition to the educational field, many public libraries are adding Internet-connected computers. But most of these computers are constantly in use with long waiting lines and frustrated patrons. Some libraries, to provide relief for the wired computers, are providing WLAN access. Library patrons can then bring in their own laptops and connect to the library's WLAN without affecting the in-house computers.

WLAN Providers

In the public-access area, specialized WLAN Internet service providers (ISPs) are emerging, much in the same way that dial-up ISPs began to appear over the past 8 to 10 years. The game plan for these WLAN ISPs is much the same as for their wired predecessors. A market clearly exists for WLAN services, and they offer that service at selected locations around the world, for a subscription fee.

If you want to connect to the WLAN at the airport or hotel, you should sign up with an appropriate service provider. Many airports, hotels, and other venues select a single provider for WLAN coverage.

Because the range of a WLAN is limited and because the provider must physically install equipment throughout the venue, it is not practical to allow every potential WLAN provider to have connectivity. However, in many multitenant locations, such as an airport, you may find a WLAN from one provider in an airline club and WLAN from another provider in a waiting area.

This can create some rather interesting opportunities for travelers, if they need to connect to one particular provider's WLAN. For example, you could be sitting in an airport gate area and have access to the wireless provider that services an adjacent airline private club, and vice versa. If you had a paid access membership to the WLAN provider that handled the club, you could probably gain access, even though you did not have a club membership. Of course, the converse is also true.

In the future, public WAN providers may create a system to allow another service's members to have access. This could work in the same way that you use your telephone calling card to make calls on another phone company's network. They simply pass the charges through to your card issuer who in turn bills you. Most travelers would not mind a minute-billing system as long as they could get wireless access and have one bill to pay.

Choosing a WLAN Access Provider

WLAN access providers are available to serve your WLAN needs in many public and private locations. A typical provider may specialize in certain types of locations or in certain types of business. For example, one provider may contract primarily with travel locations, such as airports and hotels. Another provider may specialize in retail locations, such as restaurants and coffee shops. Still another may concentrate on a few retail chains.

Your mission is to pick the company that best suits your needs. Ask yourself, Where do I need to be able to access the Internet on my laptop computer? You might look at marketing material or the Web sites of several potential providers. Call your favorite hotels to find out if they use a WLAN provider. You may need to contact a particular hotel in a large city to find this out. What airports do you frequently travel through? What airline clubs do you belong to? Check out their WLAN options. You can use Table 9.1 to note your favorite travel locations and compare several WLAN providers.

TABLE 9.1 Worksheet for Choosing a WLAN Provider

WLANs	Provider A	Provider B	Provider C
Airports			
Hotels			
Coffee Shops			
Other			

With this information, you can sort through the choices to determine the best provider for you. It is a good idea to pick one that covers most of your intended wireless locales. However, do not be discouraged if you cannot cover every base. It is possible to fill in the missing venues as you go.

Most WLAN providers provide several options for gaining access. In most cases, you will have to subscribe to the provider's service network. The most common plan is a flat monthly fee for unlimited use or for use to a certain level. From one perspective, this is a little like the difference in the fees between a home (or office) telephone line and a cellular telephone line. Most wire-line phone companies (the conventional local-exchange carriers) sell you local phone service for a flat monthly rate. However, most cellular phone carriers sell you a bulk-usage plan for a fixed rate and charge you by the minute if you exceed your plan.

If you get your WLAN service subscription through a company that provides service in the locations you wish to use, you will pay a flat monthly amount for access in any of those locations. If your provider covers those airports, you can go to Atlanta Hartsfield, Dallas/Ft. Worth, Chicago O'Hare, and San Jose International airports and use their wireless access LAN without paying extra.

However, if you wind up in Minneapolis, and your provider is not there, how do you connect? One option is to sign up with another provider. But if your stay is brief and your visits are infrequent, this might not be a good option. Fortunately, some providers offer daily and measured plans.

A daily payment plan allows you to use the provider's WLAN service for just that day at a greatly reduced rate. You can normally get a day-rate WLAN connection for about the price of a magazine in the airport shop, or the price of a nice single-malt Scotch...not a great one, but a nice one. So your choice is to read, gossip, drink, or surf the Internet and catch up on your work. The decision is yours.

The day rates are typically about one-sixth the cost of a monthly rate, so you can easily determine whether you would be better off going for the one-month rate. Most plans have a one-month minimum, and some have a 30-day cancellation, so you might have to compare these rates against a two-month fee.

It would be impossible to make a comprehensive and current list of WLAN service providers in this format. The field is growing rapidly, and locations and provider options are being added almost daily. For a current list of providers and special values, surf to www.BuildYourOwnWirelessLAN.com.

Coverage Areas

When you use a WLAN in a public place or a public system in a hotel or airport, you may have problems getting an adequate wireless signal. Unlike the SOHO WLAN that you learned to build in previous chapters, this environment is not under your control. You have no influence on the design of the wireless coverage, the placement of the APs, or the antennas used.

Typically, a coverage problem results from a design plan that tries to use only a minimum number of APs. The primary reason for this is to save costs. The sparse wireless backbone may not provide proper coverage throughout the entire area. Often building designs constrain the placement of APs, and there may be building walls or other structures that interfere with the wireless network's operation to every nook and cranny of the building (Figure 9.6).

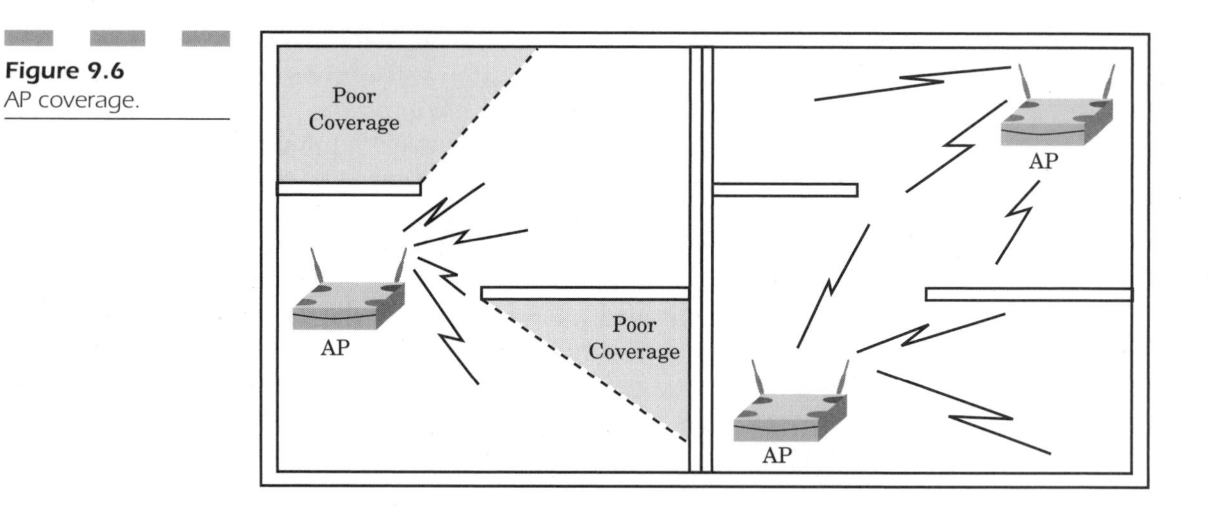

Figure 9.6
AP coverage.

Improving Range and Coverage

There are several things you can do to improve your PC's wireless performance. Even though you do not have control over the placement of the APs, you do have control over the placement and orientation of your own laptop PC. The first thing to do is to visually locate the APs, or at least their antennas. Building owners are very sensitive about appearances and interior design elements, and a wireless AP and its antenna just may not fit their idea of correct aesthetics. The WLAN access installers may be constrained in their placement of the visible elements of the WLAN.

Take a look around, particularly in the ceiling areas. You can often recognize the WLAN antennas as unusual objects hanging from the ceiling or wall. In areas such as airport terminals, a very slender antenna may be used. The small antenna typically is mounted upside-down on a ceiling tile and looks like a small plastic blade, perhaps 4 to 6 inches in length, 1 inch wide, and ½ inch thick. These antennas are similar to those used for UHF and microwave transceivers on vehicles. You can imagine this type of antenna would offer very little wind resistance on a car. Another type of antenna is a low dome or a short, flat cylinder. This antenna is typically 4 inches in diameter and about 2 inches thick.

All antennas have directional radiation patterns that influence signal strength. The blade antenna is omnidirectional. That means that it covers equally in all directions, radially from the vertical direction. The cylinder antenna is a unidirectional antenna, with a 150 to 170° pattern. It is often used on a wall to cover in a direction outward from its mounting. You will rarely see a highly directional antenna, such as a Yagi or a dish, although they may be used in instances where coverage is needed far from the antenna, and their 5 to 20° pattern is acceptable. For more information on styles of antennas, consult Chapters 2 and 8.

As you may recall from earlier chapters, WLAN signals are extremely line of sight. If you cannot see the antenna, it surely cannot see you. However, in open spaces in large buildings, such as airports, metallic objects, such as structural steel, metal doors, and decorative metal sheathing, may allow the WLAN signals to bounce around obvious obstructions. Don't make the mistake of assuming that the radio waves behave exactly like visible light. Their behavior is very similar, and some good comparisons can be made, but, in absolute terms, reflection and absorption of wireless signals are very different from those of visible light.

A good rule of thumb is to try two or more locations in a public WLAN area to see where you get adequate signal strength. You can use the W-NIC management utility to test signal strength, as shown in Figure 9.7. With the utility running, turn the laptop around to see if the signal strength improves. If you think your body or your chair is in the way, move to another seat. If you can stand and walk around with the laptop utility running, you can get a good idea of the signal's hot and cold spots. Pick a location where the signal is good and you can work comfortably.

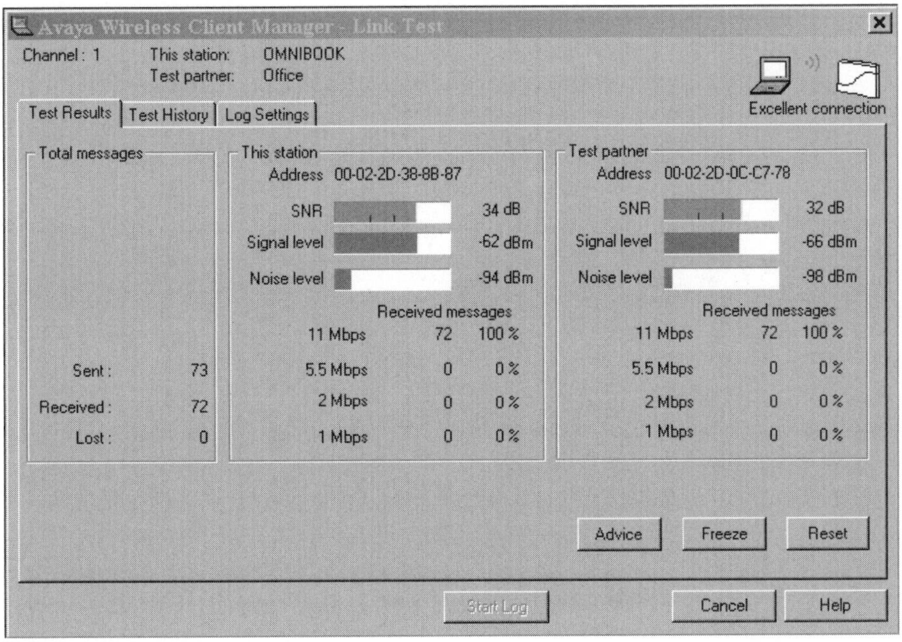

Figure 9.7 Using the link test to monitor antenna orientation.

Using External Antennas

In addition to the physical location and orientation of your wireless laptop PC, you may be able to control the W-NIC's antenna. Remember that wireless link performance depends on the power of the transmitters, the distance between transmitter and receiver, and the performance and orientation of the antennas. Each device, the AP and the W-NIC in your laptop, has a transmitter and a receiver. If you improve the performance of the antenna on either end of the link, you improve the overall signal strength across the link, as shown in Figure 9.8.

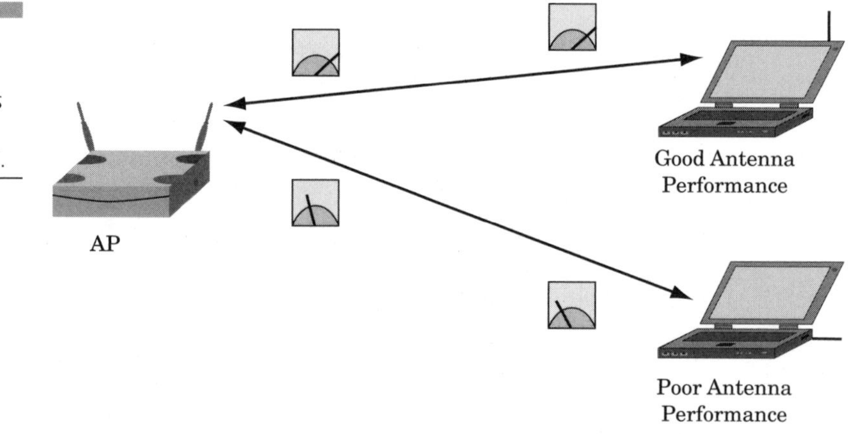

Figure 9.8 *Antenna performance affects signal strength at both AP and W-NIC.*

It makes no difference which antenna's performance you improve. In other words, a positive change in your PC's antenna is just as beneficial as a change of equal magnitude in the AP's antenna. You obviously have control over only your own PC, so let's make some positive changes there.[2] If you have a W-NIC with a fixed, built-in antenna, your only options are to reorient the antenna (with the PC attached) or physically move to a better signal area. However, if you have one of the W-NICs with the tiny moveable antenna, you should open the W-NIC manager, note the signal strength, and see if you can get a better signal by moving the antenna around. Some W-NICs have a pair of tiny antennas. On some of these units, you can select whether to use one antenna for

[2]Talking to your antenna will not help it overcome its problems. It is not a plant or an animal. It is an inanimate object that responds to the laws of physics and, perhaps, to Murphy's Law (of the Perversity of Inanimate Objects).

transmit and the other for receive or both antennas for transmit and receive.

Most of the time, wireless AP antennas are vertically polarized. This means that your PC's antenna will perform better if it matches the AP's polarization. Occasionally, a commercial antenna will be mounted with horizontal polarization. Then, naturally, your PC W-NIC antenna needs to match that for best performance.

The built-in antennas in some W-NICs are curved for size and to cancel some of the directivity. As a result, they may have a rather confused pattern. You may find that in one direction there is a hot spot where you can really get a signal boost. Or, you may find a really dead spot, called a *null*, where the signal strength goes really low. Each installation in a laptop PC will vary a little because of the design and the materials in the PC case.

A few W-NIC brands have a tiny connector on the end of the card, so you can connect an external antenna, as shown in Figure 9.9. Although the external antennas may be a little pricey, they are not at all bulky, so you could carry one along with your laptop. When you are in a really difficult signal area, these external antennas can really extend your range. You may be able to operate satisfactorily in locations that yielded unacceptable low signal strengths without the antenna.

Figure 9.9
A W-NIC with an external antenna connector.

Security Risks

We covered wireless security issues at length in Chapter 6. However, there are several things that bear pointing out in regard to security in publicly accessible WLANs. Unlike our previous considerations, a public WLAN is intended to be accessed by anyone and everyone. At least it is intended to be accessed by anyone who has a subscription to the service. If someone else gains access, it is really not your concern. That is the service provider's problem. What is of your concern is whether you suffer any security risks by logging into a public WLAN.

The short answer is, yes, you certainly do have a substantial security risk on a public WLAN. The reason is simple and is very similar to the problems with cable modems.[3] Simply stated, in a public location and a publicly accessible AP, you are allowing every WLAN user within range of that AP to listen in on your conversation. Granted, a person would have to have some special software and/or knowledge to do so, but it is a very serious security risk.

Public WLANs typically must disable WEP and encryption. As we saw in Chapter 6, the use of encryption is necessary for true wireless security, and WEP must be used for at least a minimal level of privacy. But these two features add a lot of complexity to a public WLAN and are a great cause of problems. Many, if not most, of the WLAN access providers disable all security, except the network name (SSID). That means that all of your communication with the public AP, both transmissions and receptions, are "in the clear" and readable by anyone with a little common sense and a couple of readily available computer programs. What a comforting thought!

There are a couple of ways around this problem. First, if you use a VPN everything you send or receive after you create the secure tunnel will be totally secure. It will be so much unintelligible garbage to anyone who can intercept it. But a VPN by itself is no guarantee against snooping.

One of the other problems that exist on the public AP, as with the cable modem, is the ability to access your computer through Windows Networking. Anyone who also has Windows Networking and can determine your workgroup name can potentially gain access to your computer and all of your files.

---

[3]See the Tech Tip on How Cable Modems Work in Chapter 4.

There are several things you can do to mitigate this exposure. First, you can use the techniques to customize and restrict your workgroup access. This was covered in Chapter 4. It would be best to disable WINS, turn off File and Printer Sharing, and eliminate NetBIOS over TCP/IP and over NetBEUI. However, if you have to use a WINS server to access resources on your home office network (through the VPN), you cannot do all of these things.

Another step you can take is to run a personal firewall, as described in Chapter 6. Specifically, this can prevent direct NetBIOS snooping and will block an intruder's access to any ill-behaved TCP/IP ports on your computer. A personal firewall usually will not interfere with a VPN connection, but it will prevent someone from slipping through the back door to get into your computer and, perhaps, even through your tunnel to the corporate site.

If you make your public WLAN connection without a VPN and without encryption, you can still receive some protection from certain Web sites. Standard browser protocols[4] allow a secure connection between your browser and the Web server. This means that an eavesdropper gets nothing of benefit as long as the session stays secure. However, if you have used many secure sites, you will notice that they may mix secure and insecure pages. So, some of the information you send may not be all that private, although financial transactions and purchases usually are.

[4]HTTPS, or Hypertext Transport Protocol, Secure Socket Layer encrypts end to end through a public key infrastructure.

CHAPTER 10

Internet-enabled Cellular Phones and PDAs

Chapter Highlights

- Cellular Internet
- PDAs
- WAP-Enabled Web Sites
- Comparisons to WLANs

This chapter is about the other type of wireless Internet, which can be described as cellular wireless Internet access. From the standpoint of accessing the Internet, the fundamental difference between the two technologies is the way they connect to the resource that allows access to the Internet. A secondary difference, but a very practical one, is the relative speeds of the two wireless types. A third difference is the size and format of the devices that typically use the two technologies. These characteristics are illustrated in Figure 10.1.

The preceding portions of this book have been concerned with the operation of WLANs. As we have seen, the primary purpose of a WLAN is to interconnect digital computers into a cohesive wireless network. A WLAN, by its nature, is intended to serve a very limited, localized space. This is in keeping with the philosophy of a conventional wired LAN, which is mostly supported up to distances of only 100 m (329 feet).[1] WLANs can span from slightly less to slightly more than that distance indoors. The typical network speeds of WLANs are roughly in the same range, between 1 and 100 Mbps. WLANs can slow down 1 Mbps to extend their range and can operate at 11 Mbps or even 54 Mbps (for some of the newer hardware).

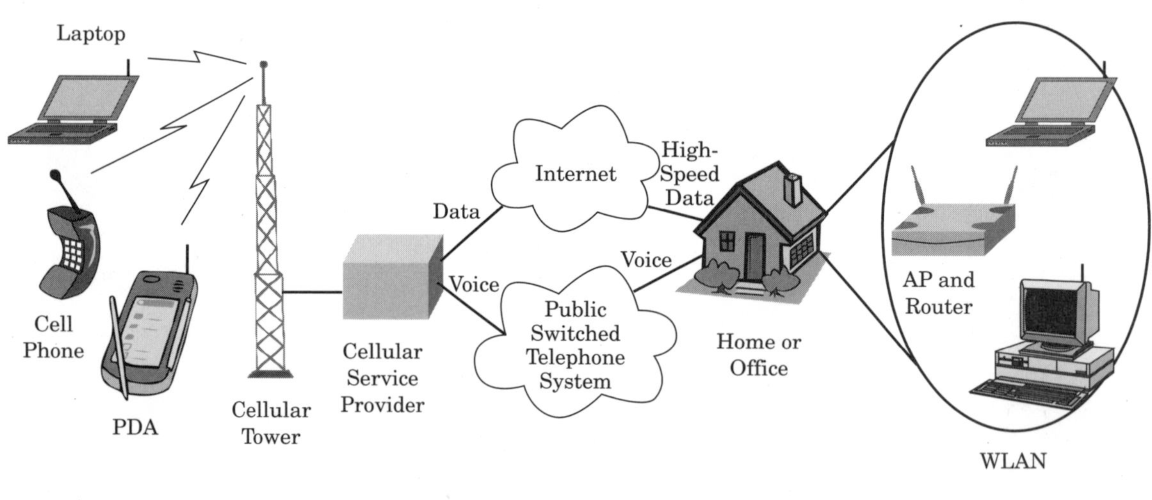

Figure 10.1 Comparison of cellular wireless Internet and WLAN Internet.

[1]Throughout this book, in keeping with the standard practice in the industry, metric conversions are approximate. In a few instances, such as this, the values for meters and feet are verbatim according to the standards (TIA/EIA-568-B and ISO/IEEE 8802.3).

In contrast, the primary purpose of the cellular network is to interconnect cellular telephones to the public switched telephone network (PSTN). That is a fancy way of saying that these networks were brought into existence to allow people to use mobile telephones—ones not connected by wire to the telephone network. However, it has long been possible to adapt mobile telephone systems to transfer data. In the modern cellular network, a secondary purpose of the network has become allowing portable cellular devices to connect to the Internet.

In the next sections, we cover some of the details of cellular network operation and the application to Internet access provided by advancements in cellular telephones and personal electronic devices. Although cellular Internet does not provide all the features of WLAN Internet access, it does provide very useful features and benefits. It is important to thoroughly understand the strengths of each technology and how you can best employ them.

Cellular Internet

Cellular Internet, as we will call this type of wireless Internet access, is made possible by the widespread availability of cellular radio. There are very few areas in most developed countries that lack a usable cellular signal. In fact, in many countries, there are overlapping, competing cellular radio systems. Often you can choose between them for best signal and features.

Originally, data transmission over mobile telephones was rather awkward. The radios had not been designed for data communications, and the first attempts to carry data used ordinary modems connected over the audio channel of the radios, as shown in Figure 10.2. This was the same way that modems were used with regular telephone lines. The typical audio bandwidth of a mobile telephone was even more restricted than a phone line.[2] Plus, variations in volume and interference had to be dealt with on the radio circuit. The data speeds were dismal (only 300 to 9600 bps) in comparison with normal telephone modem speeds (19.2 to 56 kbps). The audio inputs and outputs of the radios had to be modified to accept the modem signal. Very few manufacturers provided modem jacks on their radios, so the availability of data service was very limited. Ultimately, the data had to be connected through an ordinary mobile

[2] 3 kHz rather than 4 kHz.

phone call to a landline number (on the PSTN). The situation changed little with the introduction of analog cellular radio, other than that more channels became available.

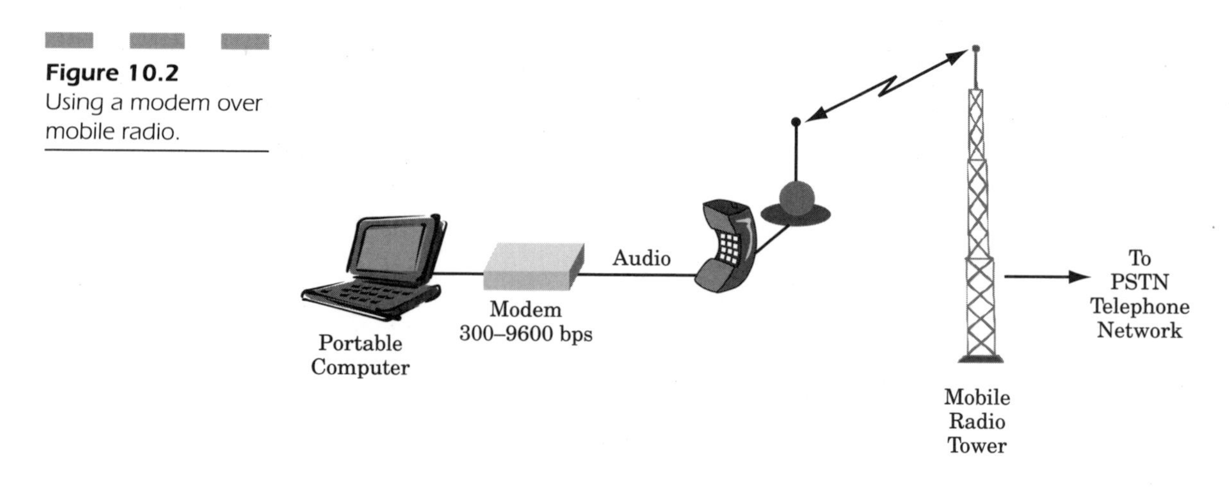

Figure 10.2 Using a modem over mobile radio.

Fortunately, the advent of digital cellular radio has changed all that. With digital cellular technology, the user's voice is converted to a digital signal and then transmitted to the cellular node (the tower), where it is reconverted to audio and placed onto the PSTN. Digital transmission has a lot of advantages, but the one of chief interest to us is that it makes data transmissions much easier and faster.

As cellular radios have become smaller and more sophisticated, it has become possible to include the two-way transmission of data to the phone itself. This is possible, at least in part, because cellular radios need digital processors to handle their primary function. It is relatively trivial to identify a transmission as data rather than as digitized voice. Data characters are quite compact compared with digitized voice, partly because the voice must be constantly converted, creating a constant stream of data between the cellular phone and the tower.

In a typical entry-level cell phone, it is very easy to convert some of the received signal to data characters and put them on the display. A brief message, such as a phone number, can be displayed quickly and easily. This makes it possible to send a numeric page message to virtually any digital cell phone, just as you would with a pager. In addition, most cellular providers offer a message service that can allow a short alphanumeric message to be sent to the phone. Cellular companies allow the message to be sent as an e-mail to the phone, and some have live operators who will transcribe a brief verbal message.

Cellular phones have become even more sophisticated, with larger displays and even flip-open keyboards. In some cases, the functions of a personal digital assistant (PDA) have been incorporated. Although these cellular phone models are more expensive, they offer additional new features, such as Internet access, that make them worth the money.

To understand the ways in which cellular Internet access differs from Internet access over a WLAN, we need to cover some details of cellular technology.

Cellular Systems

Cellular radio networks are innovative systems that greatly expand the number of mobile radio channels and use low-power radiotelephones to provide service to a large number of subscribers. This is in contrast to the older mobile radio setup that used high, centrally located towers to connect to a small number of mobile users with relatively high-power radios. In this older system, as few as 10 channels were originally available, and the system saturated an entire metropolitan area with as few as 200 users.

The cellular system uses a large number of towers arranged in a grid, with low-power transceivers and a large number of channels (Figure 10.3). The original complement of more than 600 analog channels has been multiplied many-fold by the move to digital. In addition, cellular phones can operate at very low transmitted power levels because of the proximity of the cellular towers. The low-power levels and the ultrahigh frequencies reduce absolute range and interference. This allows many more subscribers to access a cellular system because each user connects to one cell area, and the other cell areas are available to connect other users.

In addition, only adjacent cells must avoid sharing frequencies. Nonadjacent cells generally may reuse frequencies that are in use elsewhere. For example, in Figure 3.10, users A and C could be using exactly the same frequency channel, but B could not. This reuse of the channel resources allows a very large number of cellular subscribers to be accommodated and reduces the possibility of interference.

In addition to its low-power characteristics, cellular technology uses a type of frequency modulation that allows a stronger signal to totally eliminate weaker interfering signals. This phenomenon, called the *capture effect*, makes it possible for the cell phones to tolerate a fair amount of interference without the user being able to hear it. In analog cellular networks, you will occasionally hear two conversations at once or perhaps an

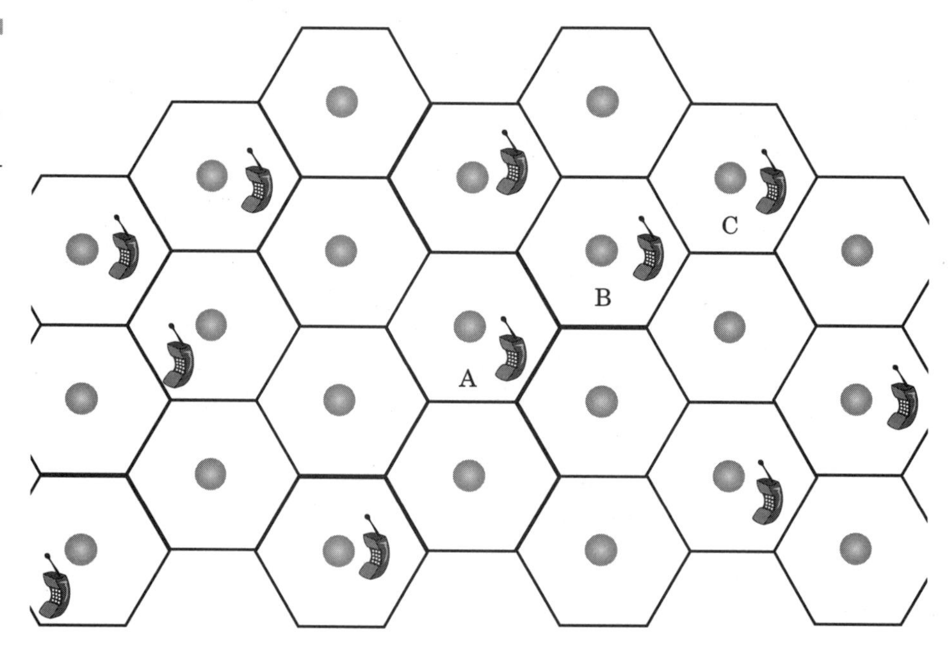

Figure 10.3 Cellular radio divides geography into a cell grid.

annoying sound from the interference. Digital phones, however, constantly check the integrity of the digital signal and do not convert bad signals to audio. Thus, with digital radio service, you may hear some occasional clipped voice or audio voids rather than the interfering signal.

Theoretically, cellular systems operate flawlessly on open terrain with ideally placed cellular towers. However, in the real world, buildings, hills, and even mountains get in the way, and it is not always possible to place a tower in the center of the grid cell. For this reason, cellular coverage may vary considerably within an area of the city or countryside. In most areas, cellular service is available from several carriers. Some cellular carriers may have great coverage in the same locations where others have very limited coverage.

Cellular Data Access

Digital cellular services are provided by a variety of service methods. In the United States, the most prominent are time division multiple access (TDMA) and code division multiple access (CDMA). In Europe and Asia, global system for mobile communication (GSM) is predominant. The digital cellular implementations are considered the second generation (or 2G) of cellular radio. GSM, however, is gaining acceptance in the United

States, and is available in many locations. In addition, an upgrade to GSM, called WCDMA (wideband CDMA), is emerging to advance to the next generation of service, or 3G.

Most data access over current digital cellular systems uses a packetized data networking technology called *cellular digital packet data* (CDPD). CDPD can use idle voice channel times to provide up to a 19.2-kbps data rate. Because of overhead, timing, and other factors, a net throughput of about 9.6 kbps is typical. However, some systems can achieve more than this.

Now that the Internet has become an important part of the cellular carrier's offering, virtually all of the digital carriers are offering data service to their cellular customers. Why wouldn't they? The technology already supports data transport, and, delightfully for the service providers, data customers spend more time online, which results in more billable minutes on the cellular network.

Not all cellular phones can handle Internet content. This capability is relatively new, so you can expect only the newer phones to have CDPD or equivalent compatibility. If this feature is important to you, you should obtain a phone that supports Internet access. Most cellular phones, even those with Internet capability, have very limited displays, so the amount of Web content you can display is limited. Further, the data rate at which Web site content can be transmitted to a cell phone is very low. The download size is critical to good performance over a cellular Internet connection. Good sites will recognize a cellular Internet user and present their content in a multilevel, indexed form. This makes it much easier on the user, who does not have to scroll around on a large Web page, trying to read through a display with a tiny window.

Many Web sites are now compatible with cellular phone requirements and provide specially abbreviated menus for cellular users or deliver their pages through a special gateway device that trims the site to fit. A special protocol, wireless application protocol (WAP), has been developed to promote greater readability and compatibility with wireless Internet devices, such as cellular phones.

Speed Considerations

Speed rules on the Internet, and cellular Internet access is no exception. That is why manufacturers, developers, and carriers are struggling to increase the speed available to the wireless Internet access market. Current speeds may be as low as 19.2 kbps, with throughput of half that.

That is probably adequate to deliver text content to cellular phones with limited displays, but normal Web pages average 50 to 100 kB. Content for these cellular devices must be constrained to short indexes or selection menus, with short forms for filling in IDs and passwords.

Table 10.1 shows the relative effects of data speed in delivering content. As you can see, moderate Web pages take quite a long time to completely download if the page size is very large. A brief menu, which could be represented by the text, "Trade Stocks" "(1) Buy Stocks" "(2) Sell Stocks" "(3) Option Chain" "(4) Charts", can be contained in about 100 B, and could download to your phone in about one-tenth of a second. In contrast, a standard web page of 30 to 60 kB would take nearly a minute to download, but the wait would seem interminable.

TABLE 10.1
Download Speed versus Web Page Size

Link Speed*	100-B Page	1-kB Page	10-kB Page	50-kB Page
9.6 kbps	0.1 sec	1 sec	10 sec	52 sec
19.2 kbps	0.05 sec	0.5 sec	5 sec	26 sec
56 kbps	0.02 sec	0.2 sec	2 sec	9 sec
128 kbps	0.01 sec	0.1 sec	1 sec	4 sec
800 kbps	0.001 sec	0.01 sec	0.1 sec	0.6 sec

* *Load speeds assume a 2-bit overhead per byte. Results are rounded.*

Advanced cellular phones are being offered with larger screen capacities. Some of these units have displays that mimic those of the palmtop computers.[3] Because the prospect of holding a small computer up to your head to talk on the wireless phone is daunting, manufacturers are offering cell phones that clamshell open to reveal a fairly large display and a full miniature keyboard. Some of the new cell phone–Internet phone combos offer color displays and voice commands.

What all this means is that the needed data rates are going to increase. The modest text rate that was fine for the 4-line × 20-character display is far too slow for a 640 × 480 pixel color-graphic screen with a fairly complete Web browser. Too slow, that is, unless you are willing

[3] *Palmtop* is another term for personal digital assistant. PDAs can range from a simple calendar and phone-number index to fully functional, if somewhat limited, computers.

to keep looking at those 4 × 20 menus on your expensive, fully functional cellular Internet terminal.

In addition to the built-in cellular telephone applications, cellular radio adapters are available in a PC card[4] format to allow your laptop computer to connect to the Internet through the cellular network, but these adapters use the relatively slow cellular CDPD speeds. A measly 9.6 kbps is just too slow. If you ever experienced trying to browse the Internet before high-speed modems became available, you know that there is a world of difference between 9.6 and 56 kbps. Some of us thought that the World Wide Web was really the "World Wide Wait" or that the Web represented the cobwebs that grew while we waited for the page to finish loading.

By some reports, 40 to 60 percent of households and businesses in the United States will have high-speed access by 2006. High-speed access, here, is defined as the type of speeds that are achieved by cable modems, around 800 kbps or higher. If cellular Internet access is to become popular, particularly to connect laptop computers and advanced PDA/cell phones, it must offer access that is as fast as possible.

The new technology innovations, being touted as 2.5G and 3G[5] cellular networking, will offer these increased speeds and other performance-enhancing features. 3G, in particular, touts speeds of as much as one-fifth that of 11 Mbps WLANs. Such a speed increase, on the order of 2 Mbps, would certainly meet the foreseeable requirements for wireless Internet access. However, a tremendous build-out of new facilities is needed to accomplish 3G. All of the user radios—the cell phones and cellular modems—also will have to be replaced, as well.

The prospect of cellular data speeds in the range of WLAN speeds presents an interesting aspect for acquiring the consumer's favor. Along with that favor comes a lot of money in any currency. It appears that the battle has begun. With the increasing availability of WLAN connections through service aggregators and the opening of freely shared WLAN systems in major cities, you may have an interesting decision to make for your wireless Internet connection.

Screen and Keyboard Formats

The screen limitations and methods of entering characters present major challenges to Internet access from cellular phones. Most basic cellular

---

[4]PCMCIA.
[5]G stands for the generation of cellurlar technology.

phones have multiline displays of four to six lines, and 20 or so characters per line (Figure 10.4). As long as the message is brief, it is easily displayed. If the message is longer than the screen can accommodate, the user must scroll the display to view more of the message. Clearly, phones with such limited displays cannot offer much to support Web pages, but they can be used to view and respond to brief e-mail messages.

Figure 10.4
Some cellular phones offer limited displays.

Cellular phones were designed to make telephone calls. As a consequence, they are normally equipped with a very small number of buttons to allow normal numeric dialing and a few other functions. Since the advent of rotary dialing, the number selections have also been labeled with letters of the alphabet. If you look at any telephone in the United States, you will see ABC next to the 2 key, DEF next to the 3 key, and so forth. This was meant to help transition telephone users from an operator-assisted manual call system, where each telephone CO had a two-letter prefix and each subscriber had a three-, four-, or five-digit number within that office. Each exchange office was given a memorable name, selected to match the two letters in the prefix. For example, the Walnut exchange would be represented by the WA prefix, corresponding to the digits 92, which became part of the automatically completed dialing sequence.

Because telephones already have the alphabet distributed on the numeric keys, it is possible to use the numeric keys to select letters. A message can be composed by pressing each key the number of times corresponding to the position of the letter in the alphabetic label for that key. For example, you press the 2 key once for A twice for B and three times for C. Typically, the letter choice flashes up in the next position on the cellular phone's display. After a brief pause, the letter choice becomes set and the key can be reused for another letter on that same key. For example, to enter JJ you would press 5, pause, and then 5. In this manner, all of the letters of the alphabet can be chosen. For punctuation, typically you select one of the dialing keys with no letters associated (e.g., *) and then pick from a list.

This is extremely cumbersome and time consuming. Most people would agree that it is acceptable only for infrequent or emergency use. However, this is exactly the way that most people enter the alphabetic names or descriptions with their memory-dialed numbers. On the average cell phone, there is no mouse, no mouse buttons, and no keyboard. This is the only way to enter alphanumeric information.

Most Web sites that are set up for mobile phone access recognize cellular phone clients. They know that these screen size and response issues exist, and they download only very short numbered selection lists to the phones. The user can press 1, 2, 3, or 4 to make a choice. Some sites also keep track of your preferences. If you often check the weather in Boston or the price of a particular stock, those options will be on your short list. You may have to set up these preferences manually, but you will save time in the long run.

Some phone models have canned responses that allow you to reply quickly to text messaging. This avoids the time-consuming task of punching multiple times on the tiny keys. You can often create your own stock messages to add to the list of canned responses. This functionality is similar to the operation of the simpler two-way pager models.

It is unfortunate that cell phone carriers do not offer a separate paging number with their phone service. This would let you publish a pager number that would display on your cell phone without having to carry a separate pager. Often, only a numeric message is needed to remind you to call the office or home, and it would avoid the hassle of listening to your voice mail for a simple "please call me" message. It also would allow you more privacy by not having to give out your mobile number. Anyone who carries a phone, a pager, and a PDA knows how convenient it would be to consolidate these functions.

Several models of cellular phones have been introduced to deal with the display problem. Designs are available that incorporate an abbreviated PDA screen into the phone (Figure 10.5) and others that snap open to reveal an expanded display and a small but usable full keyboard.

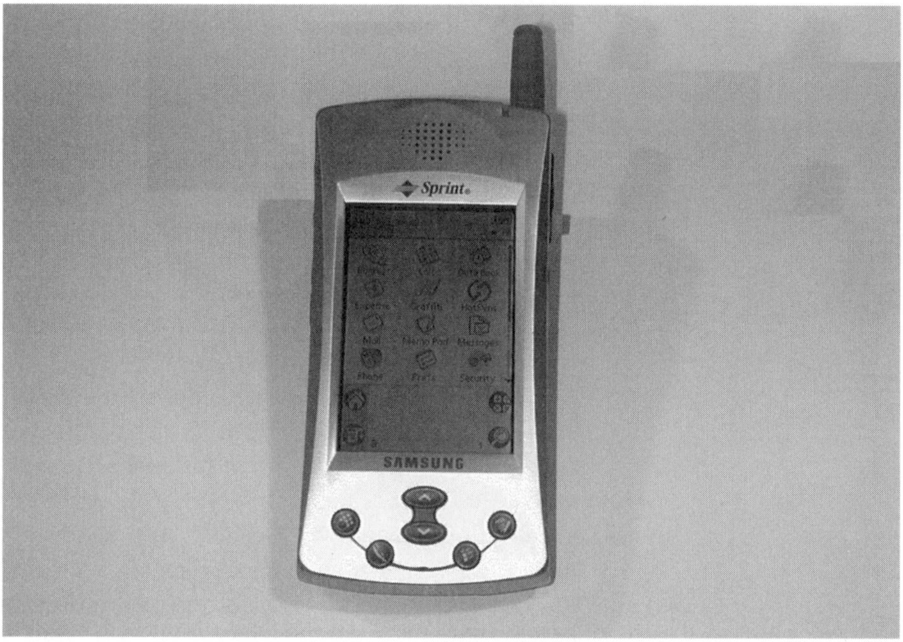

Figure 10.5
Some phones have expanded displays and keyboards.

One major problem with cellular telephones is the lack of a normal keyboard. Some of the more advanced two-way pagers have solved this problem with a small array of alphanumeric keys arranged in a standard keyboard pattern. The pager can be held in such a way that you can type one key at a time by using your thumbs. Although this is a little awkward, all computer keyboard users are already very familiar with the key layout, so it is easy to find the correct keys to compose your message.

This concept has been adapted to cellular telephones. It is still not possible to touch type (at least at any reasonable speed), and the keyboards are not perfect, but they are certainly a step or two above the multiple key-press method of the standard phone.

We are certainly close to having wireless accessories for our wireless accessories. With the introduction of cordless links, such as Bluetooth, we could theoretically link an accessory to our cell phone. It would be great to have a compact, conventional keyboard and mouse that would wirelessly connect to our cellular telephone. But then, we would almost have a small laptop computer in several sections. It seems easier just to get the laptop and the cellular network connected.

Cellular telephone models that include a built-in PDA screen offer another method of character input. As with all PDAs, these phones have a touch screen. The touch screen overlays an expanded display and is activated with a special stylus. The stylus is designed not to harm or mark the screen. Alphabetic characters can be entered by tracing the letter out on the screen, sometimes in a special area. Although this method is slower than typing on a full-sized keyboard, it is often faster than other cellular options, particularly the multiple keystroke entry method on phones that have only a dialing keypad.

Personal Digital Assistants (PDAs)

PDAs (Figure 10.6) have become an important personal productivity tool for many business people. The category name of PDA came about because of the difficulty in describing a device whose function was to automate certain manual tasks such as a calendar, calculator, and phone book, without being a full-blown computer. Typical PDAs have all of these functions, plus a structured database and a to-do list. Additional applications, such as maps, city guides, off-line e-mail, and expense reporting, are available. Add-on modules are available for most PDAs to perform functions such as global positioning system, WLAN, modem, and cellular Internet access.

An obvious value to the PDA is the ability to maintain synchronized productivity applications, such as phone book and calendar, on the portable PDA and your primary office computer. The office computer can even maintain multiple schedules and separately synchronize the PDAs of each individual. To synchronize a PDA, you simply connect it through a recharging cradle or a sync cable that is connected to your computer.

PDAs can also transmit and receive contact information and other items by using an infrared port. To send a business contact, you simply point the infrared port at another PDA or compatible device, and the transfer is made, almost instantaneously. It sure beats printing out the information on the touch screen.

PDAs are now much more than simple convenience tools. They have many of the attributes of earlier personal computers in terms of memo-

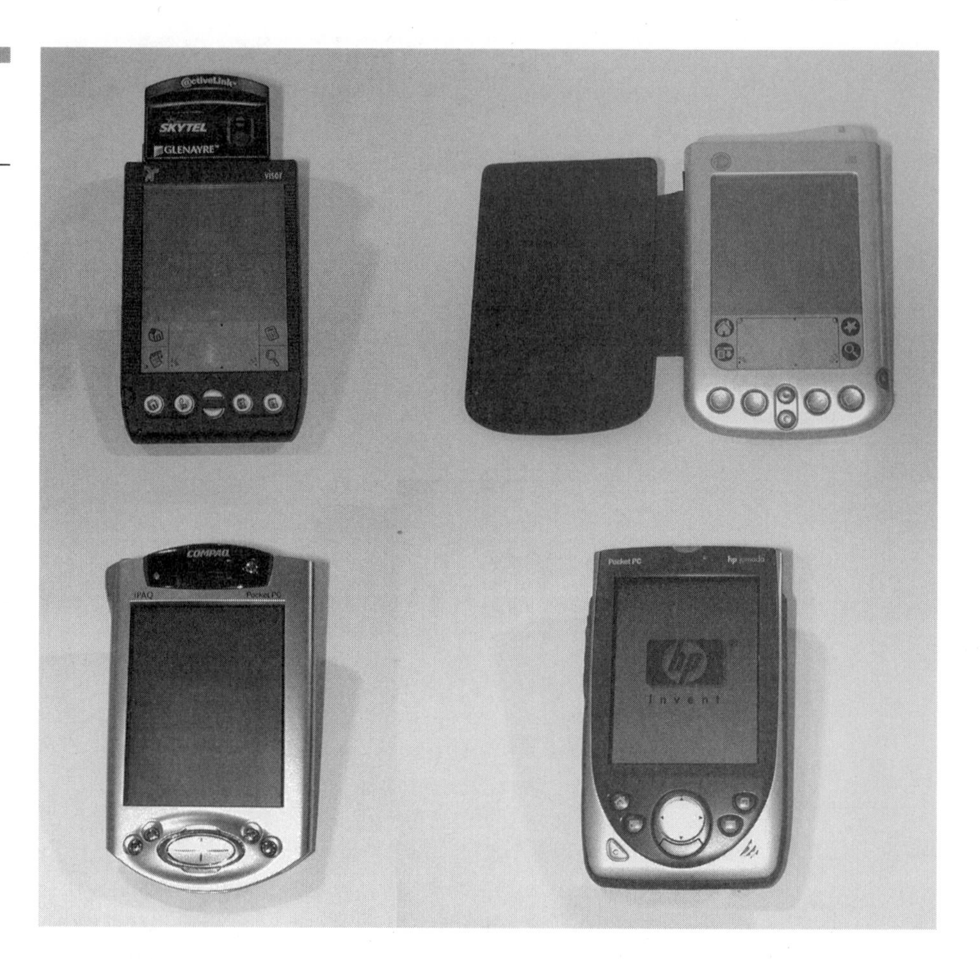

Figure 10.6
Personal Digital Assistants (PDAs).

ry and speed. In many ways, they exceed the attributes of those early systems. PDAs are indeed small computers with data, applications, OSs, and I/O. Optional fold-up keyboards, tiny printers, and network interfaces make these devices miniature counterparts of the full-size systems we use everyday.

PDAs can now offer Internet capability, with surprising functionality. The potential for their use is so vast, that an entire Web interface has been built to deliver content suitable for the PDA. This language, WAP, opens a whole new frontier for the Web. However, unlike the visions of an oppressive Big Brother in the popular novel *1984*, the PDA has turned into a really useful Little Brother. It is just as ever present, but its friendly interface and powerful features make it a tool for simplifying and empowering our business and personal lives.

Palm and Win-CE Platforms

PDAs are designed to be held in one hand, while making selections or writing with the stylus with the other hand. Essentially, they are held in the palm of the hand, which gives rise to the term *palmtop*. It was inevitable that, in the age of desktops and laptops, palmtops would also emerge. The name can partly be attributed to one of the primary developers of the innovative technology, Palm, Inc. Palm, Inc. has licensed its OS to several other companies (including cellular phone manufacturers), a very large installed customer base, and a large complement of third-party add-in programs.

A newcomer to this scene, that is really not so new, is a version of Windows, downsized for these smaller computers. Originally called Windows CE (for compact edition), the latest version of this miniature operating system is dubbed Pocket Windows. Pocket Windows (and CE) PDAs have much the same functionality as Palm OS, but applications for the two are not generally compatible.

Both types of PDAs offer a variety of platforms from several manufacturers. Models are available with monochrome or color displays. Screen size is usually 2.5×3 to 3×4 inches, and resolution varies from 160×160 to 640×480 (which is the old CGA standard). Given the small size of the screen, the higher resolutions are not of much value.

If you are shopping for a PDA, you will need to evaluate a host of items, from size and resolution, to type of display and available applications. Keep in mind that, although both types of PDAs can read and write the common word processing and spreadsheet formats, most of these documents are massive compared with the lesser memory capacity of most PDAs. Most PDAs store their data in flash memory, although tiny microdisk drives are available for some units. You will have to be sure that your PDA has the capacity for these documents if that is your need. One of the larger PDAs, which are really like a miniature laptop, will be more likely to handle these larger applications, if that is your need.

PDAs also feature many modular add-ons. Of particular importance to our current wireless subject are the cellular radio modules and the WLAN modules. Both types are offered to expand most brands of PDAs. Some PDAs even come with the wireless capability built in. Unlike cellular telephones, which are truly cellular creatures, PDAs have no predisposition and can be equipped with cellular or WLAN adapters.

Cellular Adapters

Cellular adapters for PDAs come in three types. For our purposes, we consider PDAs with built-in cellular modems to be the first type (Figure 10.7). If you are certain that you will be using your PDA for cellular Internet access, you should definitely consider one of the PDA models with a built-in cellular adapter. These units are clearly the most convenient to use because you only have to raise the antenna, enable the Internet function, and you are set to go. Their operation does not require installation of a separately purchased cellular adapter or software. The operation of the built-in adapter is designed to be 100 percent compatible with the operation of the PDA, so you don't have to worry about any of the issues that arise with a third-party vendor.

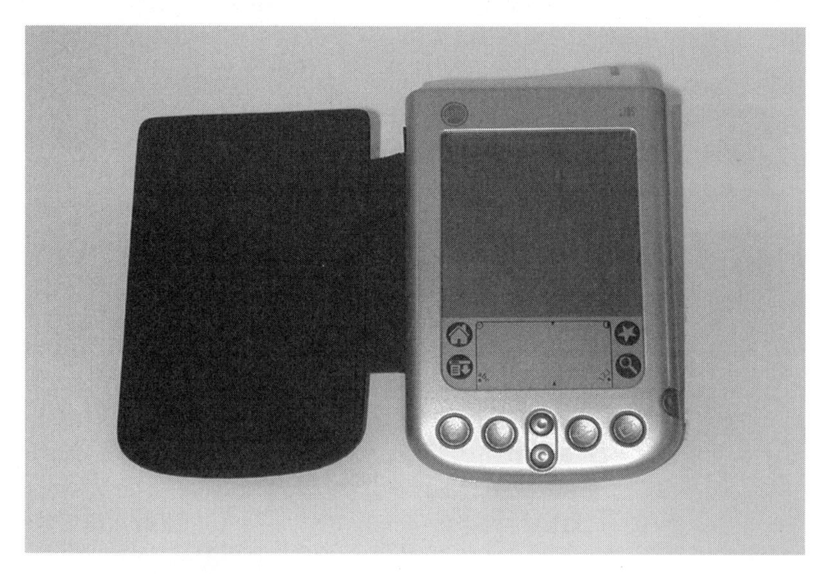

Figure 10.7
A PDA with a built-in wireless adapter.

However, as with everything, there are a few drawbacks to PDAs with built-in cellular adapters. Probably the most frequently heard comment is that they are slightly larger than their noncellular counterparts. Of course, they also cost more, because the adapter is included in the price. Another comment is that the cellular radio draws extra power, so the battery charge does not last as long. One of the advantages of the original handheld PDAs was that batteries would last for nearly 3 weeks of normal use. Most, if not all, of the cellular/PDAs depend on rechargeable batteries to handle the higher drain of the cellular radio when in use and to compensate for the shorter overall battery life. Nonrechargeable batteries would not be as practical for these units. As a result, you are a little more dependent on your recharging cradle.

Another type of cellular adapter for the PDA (Figure 10.8) comes as a module that connects through the PDA's expansion slot. Not all PDAs have expansion slots, but the more advanced models typically do. The purpose of the slot is to allow the addition of peripherals, such as external keyboards, printers, external disk drives, memory expansion, and, of course, cellular and WLAN adapters. The advantage to an expansion-slot adapter is that you don't need to carry around the extra bulk and weight of the wireless adapter when you don't want it, and there is no extra battery drain when the adapter is not attached. Of course, the adapters come with a swing-up antenna, just like the built-in adapters. Some modular adapters fit onto the top or the bottom of the PDA and essentially expand the case in that direction. Still others virtually wrap around the PDA and are mounted by connecting to the expansion slot.

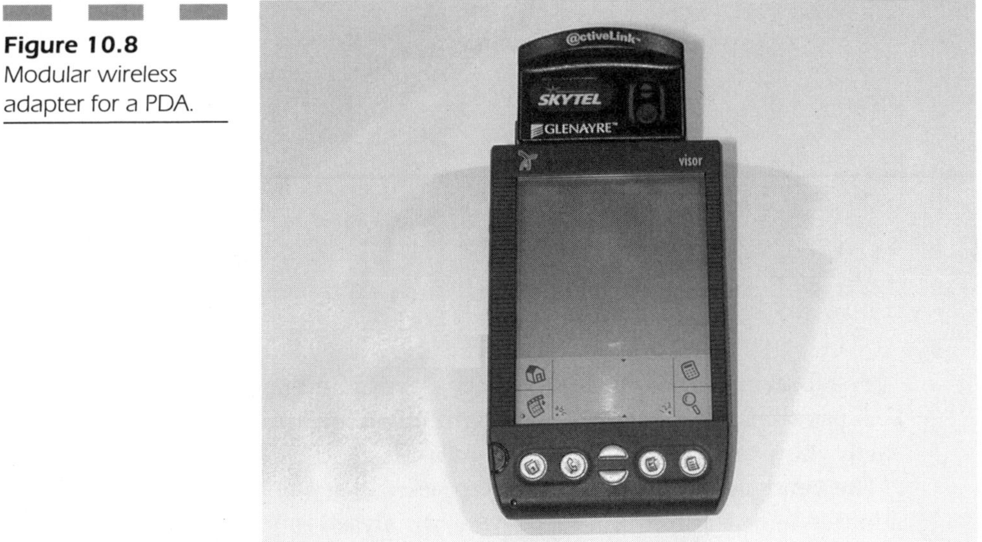

Figure 10.8 Modular wireless adapter for a PDA.

Wireless adapters on these units can be added after you have purchased the PDA, even if you originally did not have a need for a wireless connection. Unfortunately, the privilege of waiting has its costs, and the adapters are often more expensive than the feature in a built-in PDA.

The third type of wireless adapter is the PC card[6] cellular adapter (Figure 10.9). Some PDAs have a Type II PC slot, just like laptop computers. This style of adapter plugs into the specialized expansion slot, and the antenna is raised to connect. The design of a PC card adapter requires that power be supplied by the PDA, which will be a factor for battery life. However, to accommodate the PC slot, these PDAs are already a little larger than their cousins, so the cellular adapter does not add much to their bulk or weight. By the way, these same PC card cellular adapters work just fine on full-sized laptop computers, as well.

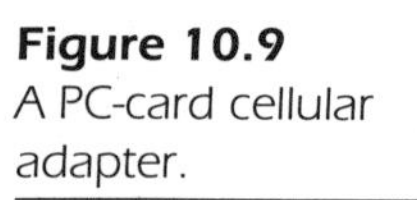

Figure 10.9
A PC-card cellular adapter.

A subscription for cellular service is assumed on a cellular telephone because that is the reason you bought one. However, PDAs do not necessarily require any cellular service at all, so you will have to add it. Most of the vendors of wireless PDA adapters will try to hide the fact that the adapter uses the cellular radio network at all—but don't be fooled (Figure 10.10). This is the only way that they can afford to provide almost universal access for the wireless appliances. Cellular adapters generally use the CDPD network, which is limited to a speed of only 19.2 kbps, but that rate should be adequate if you access Web sites that have abbreviated page support for wireless devices. Some of the newer systems operate on GSM and can offer higher data rates.

---

[6]Also called a PCMCIA card.

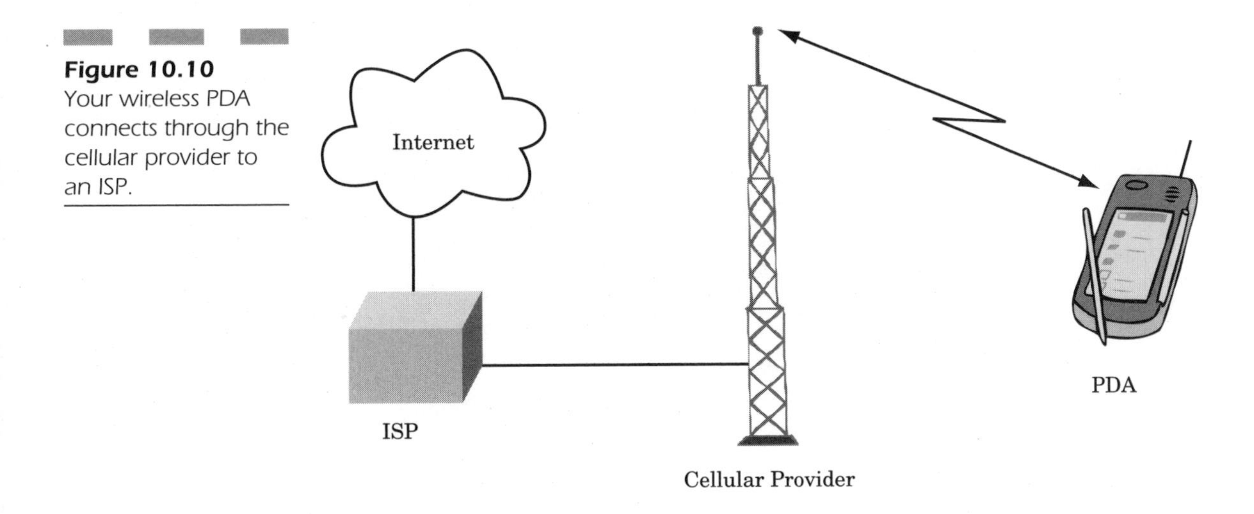

Figure 10.10 Your wireless PDA connects through the cellular provider to an ISP.

To get wireless service for your PDA cellular adapter, you simply subscribe to the monthly service. This will cost about the same as basic cellular service because you must compensate the cellular service provider and the Internet service provider, who is usually linked to your PDA cellular adapter manufacturer. However, unlike standard cellular voice service, cellular Internet normally offers an unlimited-use option.

WLAN Adapters

PDAs are not restricted to slow-speed cellular Internet access. They can be equipped with WLAN adapters (Figure 10.11), just like your laptop computer. In many ways, PDAs are a natural for roaming WLANs. It is much more likely that you would have your PDA handy in that coffee shop, restaurant, or airport than your laptop. Another advantage is that PDAs start up almost instantaneously, which is certainly more than we can say for any laptop.

If your desire is for a quick connection to the Internet to glance at your e-mail headers or check a stock, a WLAN-equipped PDA may be just the thing. With regard to Internet connectivity, WLAN links on PDAs do the same things as any other WLAN link. The same features and restrictions apply. The only thing that is different is that you will not have ordinary Windows Networking available to let you share printers and files. But when was the last time that you printed directly from your PDA anyhow? You can always synchronize your PDA to your computer to save your

Figure 10.11 Standard WLAN PC-cards can fit some PDAs.

updated contacts and calendar and print from your regular computer. The PDA can even synchronize over the Internet, if you want it to.

With public WLANs popping up all over the place, a PDA with WLAN capability makes a lot of sense. It would certainly be a desirable feature to have your schedule, your contacts, and all of your Internet needs met by one handy device. Believe it or not, some PDAs actually have a voice module that lets you also use it as a cellular telephone. What they won't think of next!

You should be aware that cellular and WLAN Internet adapters are not compatible. Each serves only its own service, and you cannot have both adapters in your PDA at the same time, unless you have a built-in adapter for one service and add the other as an expansion module. Manufacturers may offer a dual-mode cellular/WLAN adapter. Your only dilemma then will be to make sure you have picked the best service for your location and your needs.

APPENDIX

Internet Address Scheme

Internet addresses consist of 32 bits. These 32 bits are divided into 4 octets of 8 bits each and normally expressed in decimal form. For example, the binary address, 00001010 00000001 00000001 00000001 would be written as 10.1.1.1 in decimal form. Each 8-bit sequence is converted from binary to decimal form, and the 4 octets are separated with a period character.

To efficiently move data packets from the originator to the destination, the Internet addresses were subdivided into subdivisions called *subnets*. A subnet is any division of the entire 32-bit global address space. However, in practice, the divisions are made in rather orderly chunks. The global address space, from 0.0.0.0 to 255.255.255.255, is sectioned off by the octet boundries into Class A, B, and C subnets. Table A.1 below shows these subnet classes.

TABLE A.1
Subnet Classes in Internet Address Space

Class	Mask	Example Subnet	Length	Number of Nodes*
A	255.0.0.0	64.0.0.0	24-bit block	16,646,400
B	255.255.0.0	64.128.0.0	16-bit block	65,280
C	255.255.255.0	64.128.192.0	8-bit block	254

*The maximum number of nodes is 254 in a class C address because the .0 address is the address of the subnet and the .255 address is reserved for broadcasts. Smaller subnets have a similar "shrinkage" due to these two reserved addresses in each subnet.

A Class C subnet can be divided further into subnets, as can Class B and Class A subnets. A service provider typically does this for customers who want a contiguous group of addresses. The fourth octet in the subnet mask is used for this purpose.

A Class A subnet is an extremely large group of addresses, as you can see. However, when it is divided up into Class C addresses, it serves approximately 65,000 Class C customers. Each Class C subnet is a group of 255 usable addresses, but many may be unused. As a matter of fact, many of the potential addresses in the global Internet address space remain unused, which complicates the situation and has led to a shortage of remaining available addresses. To alleviate this shortage, a system of private network subnet addresses has been developed.

Private Network Addresses

A series of addresses for use in private networks has been allocated by the *Internet Assigned Numbers Authority*. These three contiguous groups of addresses may be used freely, with no further approval. These sets of addresses are available to ensure that routers do not mistakenly route internal, private network addresses as valid global addresses. WAN routers, by convention, will not route directly to or from any address within a private address group. Thus, these addresses are considered *nonroutable*. Provision is made for internal network traffic connected to the external Internet by NAT (Request for Comment [RFC-1597]). Table A.2 shows a list of the assigned private network addresses ranges.

TABLE A.2
Private Network Address Ranges

From	To	Class	Length	Number of Addresses
10.0.0.0	10.255.255.255	A	24-bit block	16,777,216
172.16.0.0	172.31.255.255	16 Bs	20-bit block	1,048,576
192.168.0.0	192.168.255.255	B	16-bit block	65,536

These address ranges should easily accommodate the very largest private network! Customarily, the 192.168.x.x block is used for most DSL and cable modem routers. However, for greater security, the Class A block might be used, with something other than the low-order network numbers shown at the beginning of the range. For example, if you used the 10.129.182.x Class C address range, it would take much more scanning to guess which address a computer was using than it would to scan a single Class C subnet. Also, frequent security targets are the more common default router and gateway addresses in a subnet, for example, .1 or .254.

Internet Request for Comment (RFC)

The development of the Internet has been a uniquely cooperative effort involving computer scientists and network engineers around the globe. In keeping with this cooperative and open environment, many of the Internet's "standards" are not really formally adopted standards in the way that most technical and government standards and regulations are approved and promulgated. Rather, the ideas that result from trying to create a well-ordered global network and the methods that are proposed to solve problems on that network are circulated throughout the Internet community as a series of technical memoranda. Each notable memorandum (the RFC) is assigned a number. The numbers become the common "handle" for the concept, sometimes in lieu of a fancy acronym, and sometimes despite it.

This process has been formalized over the years by the Internet Society (ISOC). Internet Drafts, short-term working documents, may become RFCs to provide information and guidance to the Internet community. RFC 1739, *A Primer on Internet and TCP/IP Tools*, is an example of an informational RFC. Some RFCs are placed on the standards track. These RFCs become a Proposed Standard, a Draft Standard, and eventually an Internet Standard. For example, STD 5/RFC 791 is the standard on IP.

We have mentioned only a few of the RFCs in this book. Some of these are RFC 1597, *Address Allocation for Private Networks*, which describes the use of private, nonroutable network address ranges, as we first covered in Chapter 4. Another handy one is RFC-1631, *The IP Network Address Translator (NAT)*. If you want a complete list, you can go to www.cis.ohio-state.edu/cs/Services/rfc/rfc.html.

IEEE 802.11 Channel Frequencies

The Institute of Electrical and Electronics Engineers (IEEE) standard IEEE 802.11 includes wireless operation in a variety of modes. IEEE 802.11 specifies operation in two bands, generally referred to as the 2.4-

GHz band and the 5-GHz band. The 2.4-GHz band was specified in the original "base" version of the standard. This base standard provided for two types of spread-spectrum modulation frequency-hopping spread spectrum (FHSS) and direct-sequence spread spectrum (DSSS). FHSS is an older technology that is not used in the WLAN products intended for the consumer market. The DSSS channel plan for the 2.4-GHZ band specifies a series of channels with nominal frequency assignments at 5-MHz intervals, beginning at 2412 MHz.

Different 2.4-GHz frequency allocations are provided by regulatory bodies in different parts of the world. In the United States, Canada, and parts of Europe, the operation is allowed from 2.400 to 2.4835 GHz. In Japan, 2.471 to 2.497 GHz is allowed. In Spain the range is 2.445 to 2.475 GHz. In France, the range is 2.4465 to 2.4835 GHz. The channel plan is shown in Table A.3.

TABLE A.3
IEEE 802.11 DSSS Channel Plan for 2.4 GHz

Channel	Frequency (MHz)	US & Canada	Europe (ETSI)	Spain	France	Japan (MKK)
1	2412	X	X			
2	2417	X	X			
3	2422	X	X			
4	2427	X	X			
5	2432	X	X			
6	2437	X	X			
7	2442	X	X			
8	2447	X	X			
9	2452	X	X			
10	2457	X	X	X	X	
11	2462	X	X	X	X	
12	2467		X		X	
13	2472		X		X	
14	2484					X

In 1999, the first supplement to the base standard was ratified as IEEE 802.11a. This standard adds operation in three 5-GHz bands: 5.15–5.25 GHz, 5.25–5.35 GHz, and 5.725–5.825 GHz. Collectively, these three bands comprise 300 MHz of total bandwidth, divided into 12 channels with 20-MHz separation. All three subbands are part of the Unlicensed National Information Infrastructure (U-NII) in the United States. Although a center frequency for each channel is specified, it merely represents the center of the range of the spread spectrum, orthogonal frequency-division multiplexing (OFDM) signal (Table A.4).

TABLE A.4
IEEE 802.11a OFDM Channel Plan for 5 GHz

Channel	Center Frequency (MHz)	Band and US Designation
36	5180	U-NII lower band, 5.15–5.25 GHZ
40	5200	"
44	5220	"
48	5240	"
52	5260	U-NII middle band, 5.25–5.35 GHZ
56	5280	"
60	5300	"
64	5320	"
149	5745	U-NII upper band, 5.725–5.825 GHZ
153	5765	"
157	5785	"
161	5805	"

Note that the channel numbers are rather oddly distributed, from 36 to 64 and from 149 to 161, at every fourth channel number. This distribution is due to the fact that the channels are continuously numbered in 5-MHz intervals, beginning at 5 GHz, so channel 1 is at 5005 MHz in this scheme. This method of numbering recognizes that the allocations for operation in the 5–6-GHz range may increase in the future and that a compatible method of numbering channels will exist as the blanks are filled in.

IEEE 802.11b is the second supplement to the base standard. It was also ratified in 1999 and specifies operation up to 11 Mbps using DSSS in the 2.4-GHz band. The channels do not change, although the standard recommends increasing the separation of channels in adjacent areas. A three-channel set and a six-channel set are recommended at channels 1, 6, and 11 and at channels 1, 3, 5, 7, 9, and 11, respectively. Maintaining this channel separation will reduce adjacent channel interference caused by overlap in the spectrum envelope.

IEEE 802.11g is another supplement to the standard. This supplement specifies higher speed operation in the 2.4-GHz and 5-GHz bands. Speeds range from 6 to 54 Mbps, although a popular capability of 22 Mbps is often touted because much of the existing equipment is capable of this rate without complete replacement of the W-NIC hardware components.

Cable Loss Values

In the microwave frequency ranges of 2.4-GHz and 5-GHz WLANs, the signal loss of antenna cables is very significant. Table A.5 shows losses per meter of different types of cables that are available from one manufacturer. Table A.6 shows losses at 1, 10, and 20 m, which are typical cable lengths for antennas. The relative losses for RG-58 A/U cable are shown. RG-58 A/U is commonly used for radio communication at VHF and UHF frequencies (that is, in the 150–175-MHz and 440–500-MHz frequency ranges). As you can see, the RG-58 A/U cable has much higher losses than the more sophisticated cables used for microwave LAN frequencies.

TABLE A.5

Typical Cable Loss Values per Meter

Cable Type	Radius (inches)	Bend Radius (inches)	Loss/m dB @ 2.4 GHz	Loss/m dB @ 5.8 GHz
Comscope™ low-loss coax				
LMR 400	0.405	1.0	0.220	0.355
LMR 600	0.590	1.5	0.144	0.238
LMR 900	0.870	3.0	0.097	0.161
LMR 1200	1.200	6.5	0.073	n/a
LMR 1700	1.67	13.5	0.056	n/a
Heliax™				
LDF2-50	0.375	1.6	0.190	0.323
LDF4-50A	0.500	2.0	0.128	0.217
LDF4.5-50A	0.625	n/a	0.090	0.156
LDF5-50A	0.875	3.6	0.073	n/a
LDF6-50	1.25	5.9	0.053	n/a
RG58 A/U				
Typical			1.02	1.80*

**Estimated.*
Values courtesy of WiLAN. n/a, not available.

TABLE A.6

Cable Losses for Typical Antenna Runs

	2.4 GHz			5.8 GHz		
	1 m (dB)	**10 m (dB)**	**20 m (dB)**	**1 m (dB)**	**10 m (dB)**	**20 m (dB)**
Low-loss coax						
LMR 400	0.220	2.20	4.40	0.355	3.55	7.10
LMR 600	0.144	1.44	1.88	0.238	2.38	4.76
LMR 900	0.097	0.97	1.94	0.161	1.61	3.22
LMR 1200	0.073	0.73	1.46	n/a	n/a	n/a
LMR 1700	0.056	0.56	1.12	n/a	n/a	n/a
Heliax						
LDF2-50	0.190	1.90	3.80	0.323	3.23	6.46
LDF4-50A	0.128	1.28	2.56	0.217	2.17	2.34
LDF4.5-50A	0.090	0.90	1.80	0.156	1.56	3.12
LDF5-50A	0.073	0.73	1.46	n/a	n/a	n/a
LDF6-50	0.053	0.53	1.06	n/a	n/a	n/a
RG58 A/U	1.02	10.20	20.40	1.80*	18.00*	36.00*

**Estimated.*
Values courtesy of WiLAN. n/a, not available.

Path Loss Calculations

Complete path loss calculations are rarely done for WLANs because the major application for the WLAN is for indoor (office or home) wireless networking. The reality of the situation is that unpredictable variables such as the transmission losses of walls, floors, and ceilings are not known, nor are they really predictable. In addition, other objects in the office environment, such as metal filing cabinets, desks, and even air ducts, can have an inadvertent beneficial effect on the path characteristics. Thus, for indoor use, we usually stick by the simple rules of thumb that were covered in Chapter 2 for indoor range. In specific circumstances, we can use simple field measurements to verify our rough predictions.

In contrast, for outdoor WLAN coverage and point-to-point or point-to-multipoint links, a path loss calculation is of great benefit to our planning. To do this calculation, you simply need to know the minimum (worst case) output power level and the receiver sensitivities of the W-NICs and RF components, such as antennas and cables. You can then calculate the margin by which your expected receive signal will exceed the receiver sensitivity. This margin is often called the *fade margin* because it is the amount by which your signal can fade without causing the link to fail.

Receiver sensitivities of WLAN components are given by the IEEE 802.11 standard, with supplements. For example, the minimum receiver sensitivity to support 11 Mbps is –76 dBm. If a 2-Mbps link is acceptable, you may use a signal strength of –80 dBm. (In the topsy-turvy world of negative decibels, –76 dBm is 4 dB stronger than –80 dBm.) This is the signal level necessary for a bit error rate of 1×10^{-2}, which is one error per 100 bits. This is a rather high level because an error rate of 1×10^{-3} is the minimum considered acceptable for a marginal Ethernet link. For that reason, you may want to always stay well above a fade margin of zero.

The following items can be combined to obtain a fade margin:

Transmitted power	(P_t)
Transmitting antenna gain	$(G_{a\text{-}t})$
Transmitting cable loss	$(L_{c\text{-}t})$
Receiving antenna gain	$(G_{a\text{-}r})$
Receiving cable loss	$(L_{c\text{-}r})$
Path loss	(L_p)
Atmospheric loss	(L_a)
Effective area loss	(L_e)
Antenna constant	(K_a)

Path loss occurs each way, but the terms for transmitting and receiving path ends are interchangeable because the transmissions must go in both directions. The cable losses should include the connector losses, if you use the cable loss tables in this Appendix. A good rule of thumb is to allow 0.2 to 0.5 dB per connector end. So, if two connectors are joined, the loss of each connector must be considered. To be conservative, add a 1-dB loss for the connectors at each end of the path (total of 2 dB). The antenna constant is a combination of factors, and the following example is based on a nominal antenna operating at 2.4 GHz. The fade margin is:

$$P_r - R_{sens}$$

where P_r is the received power level and R_{sens} is the desired receiver minimum sensitivity, a known value. To get the received power:

$$P_r = P_t + G_{a\text{-}t} - L_{c\text{-}t} - L_p - L_a - L_e + K_a + G_{a\text{-}r} - L_{c\text{-}r}$$

The calculation for fade margin is:

$$P_r - R_{sens} = P_t + G_{a\text{-}t} - L_{c\text{-}t} - L_p - L_a - L_e + K_a + G_{a\text{-}r} - L_{c\text{-}r} - R_{sens}$$

All of this math can give you quite a headache, so we have provided a quick and easy path loss calculator on the Web site:

www.buildyourownwirelesslan.com

INDEX

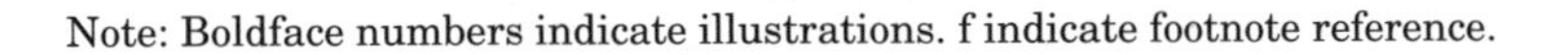

Note: Boldface numbers indicate illustrations. f indicate footnote reference.

ABOUT THE AUTHOR

JAMES TRULOVE is a networking consultant and author with over 25 years of experience with networking and technology companies, including Alteon Web Systems, Nortel Networks, Lucent Technology, Cicso, Motorola, and Intel. In his consulting work, he designs and installs wired and wireless LANs for corporate clients and advises clients on network and Internet infrastructure. His classic reference *LAN Wiring: An Illustrated Network Cabling Guide* (McGraw-Hill 2000) is now available in revised edition.